Current Topics in Microbiology and Immunology

Volume 327

Series Editors

Richard W. Compans
Emory University School of Medicine, Department of Microbiology and
Immunology, 3001 Rollins Research Center, Atlanta, GA 30322, USA

Max D. Cooper
Department of Pathology and Laboratory Medicine, Georgia Research Alliance,
Emory University, 1462 Clifton Road, Atlanta, GA 30322, USA

Tasuku Honjo
Department of Medical Chemistry, Kyoto University, Faculty of Medicine,
Yoshida, Sakyo-ku, Kyoto 606-8501, Japan

Hilary Koprowski
Thomas Jefferson University, Department of Cancer Biology, Biotechnology
Foundation Laboratories, 1020 Locust Street, Suite M85 JAH, Philadelphia,
PA 19107-6799, USA

Fritz Melchers
Biozentrum, Department of Cell Biology, University of Basel, Klingelbergstr.
50–70, 4056 Basel Switzerland

Michael B.A. Oldstone
Department of Neuropharmacology, Division of Virology, The Scripps Research
Institute, 10550 N. Torrey Pines, La Jolla, CA 92037, USA

Sjur Olsnes
Department of Biochemistry, Institute for Cancer Research, The Norwegian
Radium Hospital, Montebello 0310 Oslo, Norway

Peter K. Vogt
The Scripps Research Institute, Dept. of Molecular & Exp. Medicine, Division of
Oncovirology, 10550 N. Torrey Pines. BCC-239, La Jolla, CA 92037, USA

Marianne Manchester • Nicole F. Steinmetz
Editors

Viruses and Nanotechnology

Springer

Editors

Marianne Manchester
Department of Cell Biology
Center for Integrative Molecular
 Biosciences
Scripps Research Institute
CB262
10550 N. Torrey Pines Road
La Jolla, CA 92037
USA
marim@scripps.edu

Nicole F. Steinmetz
Department of Cell Biology
Center for Integrative Molecular
 Biosciences
Scripps Research Institute
CB262
10550 N. Torrey Pines Road
La Jolla, CA 92037
USA
nicoles@scripps.edu

Cover legend: Atomic model of 31 nm cowpea mosaic virus (CPMV) nanoparticles derivatized with gold on surface cysteines. A mutant of CPMV bearing 60 surface cysteine residues was conjugated to nanogold. Golden spheres indicating electron density of the attached gold particles are superimposed on the atomic structure of the virus capsid proteins, indicated by red, green, and purple ribbon structures. Model courtesy of Dr. John E. Johnson, Scripps Research Institute, La Jolla, CA, USA.

ISBN 978-3-540-69376-5   e-ISBN 978-3-540-69379-6
DOI 10.1007/978-3-540-69379-6

Current Topics in Microbiology and Immunology ISSN 0070-217x

Library of Congress Catalog Number: 2008931406

© 2009 Springer-Verlag Berlin Heidelberg

This work is subject to copyright. All rights reserved, whether the whole or part of the material is concerned, specifically the rights of translation, reprinting, reuse of illustrations, recitation, broadcasting, reproduction on microfilm or in any other way, and storage in data banks. Duplication of this publication or parts thereof is permitted only under the provisions of the German Copyright Law of September, 9, 1965, in its current version, and permission for use must always be obtained from Springer-Verlag. Violations are liable for prosecution under the German Copyright Law.

The use of general descriptive names, registered names, trademarks, etc. in this publication does not imply, even in the absence of a specific statement, that such names are exempt from the relevant protective laws and regulations and therefore free for general use.

Product liability: The publisher cannot guarantee the accuracy of any information about dosage and application contained in this book. In every individual case the user must check such information by consulting the relevant literature.

Cover design: WMXDesign GmbH, Heidelberg, Germany

Printed on acid-free paper

9 8 7 6 5 4 3 2 1

springer.com

Preface

Nanotechnology is a collective term describing a broad range of relatively novel topics. Scale is the main unifying theme, with nanotechnology being concerned with matter on the nanometer scale. A quintessential tenet of nanotechnology is the precise self-assembly of nanometer-sized components into ordered devices. Nanotechnology seeks to mimic what nature has achieved, with precision at the nanometer level down to the atomic level.

Nanobiotechnology, a division of nanotechnology, involves the exploitation of biomaterials, devices or methodologies in the nanoscale. In recent years a set of biomolecules has been studied and utilized. Virus particles are natural nanomaterials and have recently received attention for their tremendous potential in this field.

The extensive study of viruses as pathogens has yielded detailed knowledge about their biological, genetic, and physical properties. Bacterial viruses (bacteriophages), plant and animal eukaryotic viruses, and viruses of archaea have all been characterized in this manner. The knowledge of their replicative cycles allows manipulation and tailoring of particles, relying on the principles of self-assembly in infected hosts to build the base materials. The atomic resolution of the virion structure reveals ways in which to tailor particles for higher-order functions and assemblies.

Viral nanoparticles (VNPs) serve as excellent nano-building blocks for materials design and fabrication. The main advantages are their nanometer-range size, the propensity to self-assemble into monodisperse nanoparticles of discrete shape and size, the high degree of symmetry and polyvalence, the relative ease of producing large quantities, the exceptional stability and robustness, biocompatibility, and bioavailability. Last but not least, the particles present programmable units, which can be modified by either genetic modification or chemical bioconjugation methods.

Viruses have been utilized as scaffolds for the site-directed assembly and nucleation of organic and inorganic materials, for the selective attachment and presentation of chemical and biological moieties for in vivo applications, as well as building blocks for the construction of 1D, 2D, and 3D arrays. Here we have been fortunate to assemble a volume containing contributions by the leaders in the field, one that is marked as much by collegiality and good humor as it is by excellent science.

The chapters by E. Strable and M.G. Finn and by N.F. Steinmetz et al. address the fundamental means for performing chemistry on virion substrates and multilayered arrays. N.G. Portney et al. expand on this theme by generating hybrid virus-particle networks. The chapter by M.L. Flenniken et al. addresses the use of virus-like protein cages to generate novel materials that can be used for biomedical applications, and G. Destito et al. carry on this theme by describing the use of plant and insect viruses for biomedical imaging and vaccine purposes. Finally, P. Singh discusses harnessing the inherent tumor-targeting properties of certain viruses to achieve specificity in vivo.

Together, viruses harbor so many natural features that may be exploited for nanobiosciences. To date, it has not been feasible to synthetically create nanoparticles of comparable beauty and utility. Now there exists an unprecedented opportunity to capitalize on the vast knowledge of these VNPs and their material properties.

La Jolla, California, 2008

Marianne Manchester
Nicole F. Steinmetz

Contents

Chemical Modification of Viruses and Virus-Like Particles 1
E. Strable, M.G. Finn

Structure-Based Engineering of an Icosahedral Virus for Nanomedicine and Nanotechnology 23
N.F. Steinmetz, T. Lin, G.P. Lomonossoff, J.E. Johnson

Hybrid Assembly of CPMV Viruses and Surface Characteristics of Different Mutants 59
N.G. Portney, G. Destito, M. Manchester, M. Ozkan

A Library of Protein Cage Architectures as Nanomaterials 71
M.L. Flenniken, M. Uchida, L.O. Liepold, S. Kang,
M.J. Young, T. Douglas

Biomedical Nanotechnology Using Virus-Based Nanoparticles 95
G. Destito, A. Schneemann, M. Manchester

Tumor Targeting Using Canine Parvovirus Nanoparticles 123
P. Singh

Index 143

Contributors

G. Destito
Kirin Pharma USA, Inc., 9420 Athena Circle, La Jolla, CA 92037

Dipartimento di Medicina Sperimentale e Clinica, Universita degli Studi Magna Graecia di Catanzaro, Viale Europa, Campus Universitario di Germaneto, 88100 Catanzaro, Italy

T. Douglas
Montana State University, Dept. of Chemistry and Biochemistry, 108 Gaines Hall, PO Box 173400, Bozeman, MT 59717, USA

M.G. Finn
CB248, The Scripps Research Institute, 10550 N. Torrey Pines Rd., La Jolla, CA 92037, USA
mgfinn@scripps.edu

M.L. Flenniken
University of California, San Francisco, Microbiology and Immunology Department, 600 16th Street, Genentech Hall S576, Box 2280, San Francisco, CA 94158–2517, USA
michelle.flenniken@ucsf.edu

J.E. Johnson
Department of Molecular Biology, The Scripps Research Institute, 10550 North Torrey Pines Road, La Jolla, CA 92037, USA

S. Kang
Montana State University, Dept. of Chemistry and Biochemistry, 108 Gaines Hall, PO Box 173400, Bozeman, MT 59717, USA

L.O. Liepold
Montana State University, Dept. of Chemistry and Biochemistry, 108 Gaines Hall, PO Box 173400, Bozeman, MT 59717, USA

T. Lin
School of Life Sciences, Xiamen University, Xiamen, Fujian, PR China

G.P. Lomonossoff
Department of Biological Chemistry, John Innes Center, Colney Lane, Norwich
NR4 7UH, UK

M. Manchester
Department of Cell Biology, Center for Integrative Molecular Biosciences,
The Scripps Research Institute, CB262, 10550 N. Torrey Pines Road, La Jolla,
CA 92037, USA
marim@scripps.edu

M. Ozkan
Department of Electrical Engineering, University of California, Riverside,
A241 Bourns Hall, Riverside, CA 92521, USA
mihri@ee.ucr.edu

N.G. Portney
Department of Bioengineering, University of California,
Riverside, A241 Bourns Hall, Riverside, CA 92521, USA

A. Schneemann
Department of Molecular Biology, Center for Integrative
Molecular Biosciences, Scripps Research Institute, CB248 10550 N.
Torrey Pines Road, La Jolla, CA 92037, USA

P. Singh
Division of Hematology and Oncology, Department of Medicine,
Building 23 (Room 436A), UCI Medical Center, 101 City Drive South,
Orange, CA 92868, USA
pratiks@uci.edu

N.F. Steinmetz
Department of Cell Biology, The Scripps Research Institute,
10550 North Torrey Pines Road, La Jolla, CA 92037, USA
nicoles@scripps.edu

E. Strable
Dynavax Technologies Corp., 2929 Seventh Street, Suite 100, Berkeley,
CA 94710-2753, USA

M. Uchida
Montana State University, Dept. of Chemistry and Biochemistry,
108 Gaines Hall, PO Box 173400, Bozeman, MT 59717, USA

M.J. Young
Montana State University, Dept. of Chemistry and Biochemistry,
108 Gaines Hall, PO Box 173400, Bozeman, MT 59717, USA

Chemical Modification of Viruses and Virus-Like Particles

E. Strable, M.G. Finn (✉)

Contents

Introduction	2
Cowpea Mosaic Virus	5
Traditional Bioconjugation Strategies	7
Tyrosine-Selective Bioconjugation Strategies	13
Copper(I)-Catalyzed Azide-Alkyne Cycloaddition	14
Conclusions	15

Abstract Protein capsids derived from viruses may be modified by methods, generated, isolated, and purified on large scales with relative ease. In recent years, methods for their chemical derivatization have been employed to broaden the properties and functions accessible to investigators desiring monodisperse, atomic-resolution structures on the nanometer scale. Here we review the reactions and methods used in these endeavors, including the modification of lysine, cysteine, and tyrosine side chains, as well as the installation of unnatural amino acids, with particular attention to the special challenges imposed by the polyvalency and size of virus-based scaffolds.

Abbreviations CCMV: Cowpea chlorotic mottle virus; CPMV: Cowpea mosaic virus; DMSO: Dimethyl sulfoxide; EDC: 1-Ethyl-3-(3-dimethyllaminopropyl)carbodiimide hydrochloride; HBA: Hepatitis B virus; HSP:Heat shock protein; MjHSP: *Methanococcus jannaschii* heat shock protein; MMPP: Magnesium monoperoxyphthalate; MRI: Magnetic resonance imaging; NHS: N-hydroxysuccinimide; NωV: *Nudaurelia capensis* ω *virus*; RNA: Ribonucleic acid; TMV: Tobacco mosaic virus; TYMV: Turnip yellow mosaic virus; UV: Ultraviolet; VNP: Viral nanoparticles; VLPVirus-like particle

M.G. Finn
CB248, The Scripps Research Institute, 10550 N. Torrey Pines Rd., La Jolla, CA 92037, USA
e-mail: mgfinn@scripps.edu

M. Manchester, N.F. Steinmetz (eds.), *Viruses and Nanotechnology*,
Current Topics in Microbiology and Immunology 327.
© Springer-Verlag Berlin Heidelberg 2009

Introduction

Chemistry is a science defined and distinguished by the linking of *structure* with *function*. Enabled by ever more sophisticated tools for determining structure and investigating function, it has through the past 150 years been applied to larger and more complicated molecules and molecular assemblies, such that the boundaries between biology and chemistry are rapidly disappearing. The astonishing modern development of molecular biology is in large measure a story of how the techniques and attitudes of chemistry have been brought to bear on biological systems. A burgeoning interest in biomaterials has also developed from this fruitful intersection of disciplines. In recent years, we and others have sought to extend the historical expansion of the chemical sciences to viruses – the largest molecular assemblies to have been structurally characterized to date, straddling the boundary between inanimate matter and life. We perceived a unique opportunity to employ viruses, which are tailorable at the genetic level, as reagents, catalysts, and scaffolds for chemical operations. While achieving these goals also requires knowledge of the fundamental aspects of virus reproduction and evolution, here we focus on the chemical manipulation of viral capsids. For these purposes, we use the terms "virus," "capsid," and "virus-like particle" interchangeably, focusing on the protein shell derived from a virus. In some cases, the infectious virion may be used, but usually the protein shell is employed without one or more essential components that would allow it to propagate in a host organism. Many of the principles and techniques discussed here also apply to other self-assembling multi-protein structures such as ferritin, heat-shock proteins, and vault proteins. The overall term "protein nanocages" is an apt label for this entire family of materials.

From the chemist's point of view, viruses are captivating for the following reasons:

1. Their size range, from approximately 10 nm to more than a micron, is unique for organic structures characterized at atomic resolution (Fig. 1). While species such as colloids and polymers of comparable dimensions (200–800 Å in diameter) may be created in the laboratory, all are amorphous.
2. Unlike other materials in this size range, viruses are often perfectly monodisperse in size and composition. Only in rare cases does any particular capsid exist in more than one size or shape.
3. They can be found in a variety of distinct shapes (most commonly icosahedrons, spheres, tubes, and helices) and with a variety of properties (such as varying sensitivities to pH, salt concentration, and temperature). If the user desires a particular nanostructural feature, it may already have been invented by nature, just waiting to be exploited by the alert chemist with access to protein expression and purification facilities and expertise.
4. They have constrained interior spaces that are accessible to small molecules but often impermeable to large ones, offering opportunities for assembly and packaging of cargoes.
5. Their composition may be controlled by manipulation of the viral genome. Nonnative oligopeptide sequences may often be introduced at solvent-exposed positions of the virus coat protein with standard mutagenesis protocols and amplified

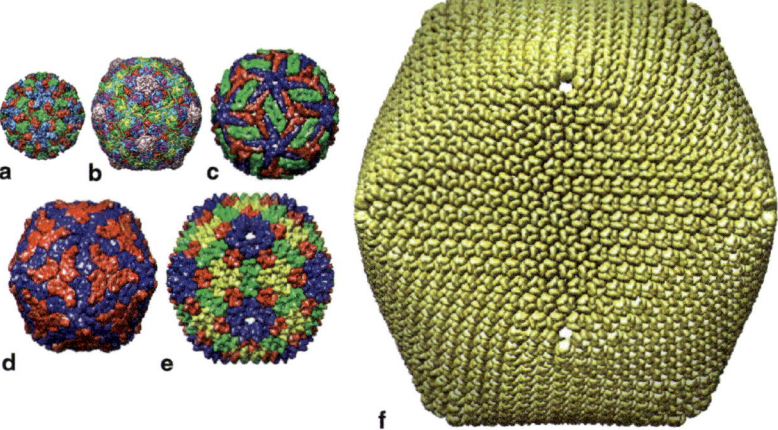

Fig. 1a–e Structurally characterized icosahedral viruses, illustrating a range of sizes (Shepherd et al. 2006). **a** Norwalk virus (Prasad et al. 1999). **b** Bacteriophage HK97 (Wikoff et al. 1999). **c** Dengue virus (Kuhn et al. 2002). **d** Rice dwarf virus (Nakagawa et al. 2003). **e** Bacteriophage PRD1 (Martin et al. 2001). **f** Paramecium *Brusaria chlorella* virus (Martin et al. 2001). The smallest has a diameter of 37 nm and the largest has a diameter of 170 nm

to an extent limited only by the efficiency of the infection or expression system. It should be noted, however, that such efforts are not as simple as they might seem. In making changes to self-assembling proteins of this kind, one must take care to leave undisturbed those regions of the landscape that are responsible for the intermolecular interactions that guide and stabilize assembly. One can increase the chances of a successful outcome by choosing sites remote from subunit interfaces, but seemingly innocuous alterations can occasionally have deleterious effects (analogous, perhaps, to allosteric effects in enzymes that can occur at great distances from active sites).

6. They represent the ultimate examples of self-assembly and polyvalence. The highly cooperative nature of capsid protein interactions makes virus particles very stable, and functional groups are displayed in multiple copies about the icosahedral spheres. Chemists may profitably think of them as very large, pre-fabricated dendrimers.

7. They can be made in quantity. Typical preparations, requiring only a few hours of effort, provides substantial yields of assembled capsids from host (plants, cultured insect cells, cultured bacterial cells) cell masses – often in the range of 0.1%–1% by weight. Most importantly from a practical perspective, viruses exhibit unique densities, making purification techniques far simpler and faster than those required for most proteins, and thus adaptable to large scale.

8. They are often more stable toward variations of pH, temperature, and solvent than standard proteins, thereby providing a wider range of conditions for their isolation, storage, and use. This property can be enhanced by using virus-like particles evolved (Flenniken et al. 2003, 2006) or designed (Ashcroft et al. 2005) to survive high-temperature settings.

9. They have large surface areas, which allow for the display of many copies of the same molecule or many different molecules, without concerns of steric crowding. Such polyvalence presents interesting opportunities for both chemical and biochemical interactions.

We describe here the first steps taken by our laboratory and others over the past several years to learn the chemical reactivity of virus capsids and to develop methods for the site-selective modification of such particles. These enabling technologies are moving rapidly, and so readers are encouraged to consult the primary literature for updates and improvements. The particles discussed are shown in Fig. 2.

The nonspecific attachment of polyethylene glycol chains to viral vectors to modify their properties of biodistribution or immunogenicity was reported by several laboratories in the 1990s (Zalipsky 1995; Chillon et al. 1998; Marlow et al. 1999; O'Riordan et al. 1999; Paillard 1999). However, the first manipulations of virus-like particles for *chemical* purposes may be found in the groundbreaking work of Mann, Douglas, Young, and coworkers, dating from the early 1990s (Meldrum et al. 1991; Douglas et al. 1995). These investigators, inspired by the natural function of the iron storage ferritin cage, made a variety of particles that lack encapsulated genetic material and have interior protein surfaces that nucleate

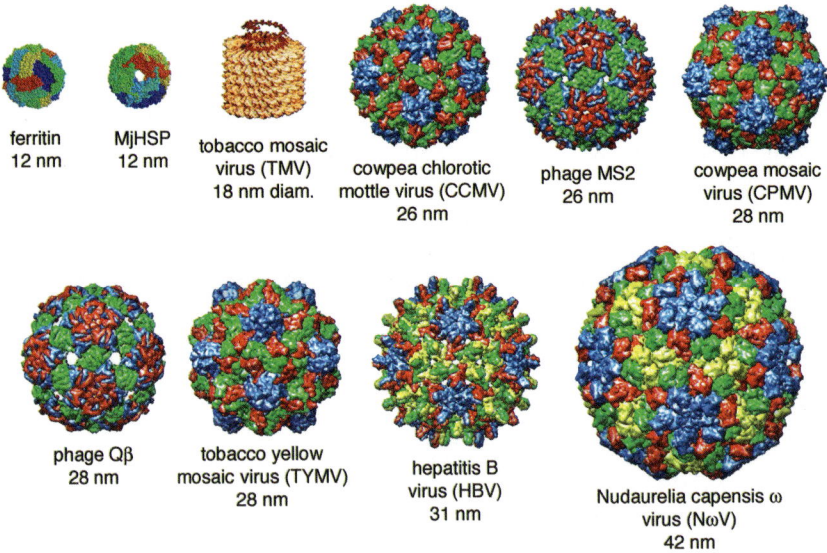

Fig. 2 Viruses and virus-like particles mentioned in this chapter, ordered by average diameter. Except for TMV, the images are colored to distinguish the symmetry-related subunits and are taken from the VIPER database (http://viperdb.scripps.edu) or the Protein Data Bank (for ferritin and MjHSP; www.rcsb.org/pdb/). Structural data comes from the following papers: ferritin (Granier et al. 1997), *Methanococcus jannaschii* heat shock protein (MjHSP) (Kim et al. 1998), TMV (Namba et al. 1985), CCMV (Speir et al. 1995), MS2 (Golmohammadi et al. 1993), CPMV (Lin et al. 1999), Qβ (golmohammadi et al. 1996), TYMV (Canady et al. 1996), HBV (Wynne et al. 1999), NωV (Munshi et al. 1998)

the size-constrained synthesis of inorganic materials (Douglas and Young 1998, 1999, 2006; Douglas 2003). The use of biological nanocages as reaction vessels and as templates for inorganic, metallic, and semiconductor materials synthesis are powerful themes, and work in these areas is certainly flowering (Klem et al. 2005a, 2005b; Radloff et al. 2005; Juhl et al. 2006; Tseng et al. 2006; Niu et al. 2006). For reasons of space, however, we restrict ourselves here to a discussion of organic chemical manipulations that make discrete covalent bonds to amino acid side chains of virus-like structures.

Cowpea Mosaic Virus

Cowpea mosaic virus (CPMV) has been the most extensively studied virus particle for purposes of polyvalent display using chemical conjugation. CPMV, the type member of the Comoviridae family, is a nonenveloped virus with a two-part single-stranded RNA genome (Table 1). Each of the two genomic RNA molecules is separately encapsidated in individual virus particles with identical (co-crystallizing) capsid structures. The RNA 1 gene product encodes the replication machinery, and is the larger of the two RNA molecules. RNA 2 encodes both the capsid and movement proteins (Lomonossoff and Johnson 1991; Stauffacher et al. 1987). Encapsidation of the two differently sized RNA molecules gives rise to particles with slightly differently densities, which can be separated using sucrose or cesium chloride gradients, and are therefore referred to as the middle and bottom components. Capsids devoid of RNA (which constitute less than 5% of natural CPMV particles) have the lightest density and are referred to as top component (Bruening and Agrawal 1967). Taking into account the relative abundance of these components, the average molecular weight of CPMV isolated from its black-eyed pea (*Vigna unguiculata*) host is 5.6×10^6 daltons. Infectious clones of both RNA 1 and RNA 2 are available, and allow site-directed mutations or peptide insertions to be made in the capsid proteins (Dessens and Lomonossoff 1993; Lomonossoff 1996; Lin et al. 1996). It is necessary to have cDNA copies of both RNA1 (pCP1) and RNA 2 (pCP2) for an infection to be produced in plants.

Table 1 Vital statistics of cowpea mosaic virus

RNA	No. of nucleotides	MW	Capsid protein	No. of amino acids	MW
RNA 1	5889	2.01×10^6	L subunit	374	41,249
RNA 2	3481	1.22×10^6	S subunit	213	23,708
Virus components					
	RNA	MW			
Top	None	3.94×10^6			
Middle	RNA 2	5.16×10^6			
Bottom	RNA 1	5.98×10^6			

The coat proteins of the CPMV capsid are produced as a fusion polypeptide that is separated by proteolytic cleavage, generating the 23-kDa small subunit and the 41-kDa large subunit. Sixty copies of both the large and small subunits come together to form an icosahedral capsid that surrounds the genomic RNA. The initial crystal structure was solved to 3.5-Å (Stauffacher et al. 1987) resolution and then later refined to 2.8-Å resolution (Lin et al. 1999). CPMV capsids have an average diameter of 30 nm, with a capsid thickness of only 12 Å. The surface topology of the CPMV capsid is characterized by protrusions at the five- and threefold axes of symmetry and a valley at the twofold axis of symmetry (Fig. 3).

The secondary structure of the CPMV capsid is dominated by nonhomologous β–sandwich domains, two in the large subunit and one in the small subunit. CPMV capsids have a pseudo T=3 surface lattice, in which each β-sandwich domain occupies the spatially equivalent position in a T=3 capsid. The single domain of the small subunit is found to cluster around the fivefold axis of symmetry, while the two domains of the large subunit are clustered at the twofold axis of symmetry (Fig. 3). This type of detailed structural information is critical for understanding how the local environment of an amino acid affects its reactivity and is the main reason that only those virus particles that have been characterized by x-ray crystallography have been chosen for chemical exploitation.

CPMV exhibits most of the other advantageous features listed above for chemistry-friendly viruses as well. Because the virus propagates efficiently in plants, scale-up is relatively easy: gram quantities of particles can be isolated from a kilogram of infected leaf tissue (Lomonossoff and Johnson 1991). The CPMV virus particles are quite stable to a wide range of pH and temperature conditions; for

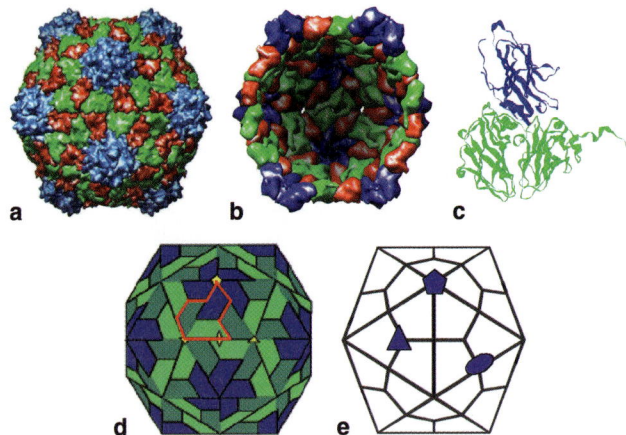

Fig. 3a–e Structure of the cowpea mosaic virus capsid (Shepherd et al. 2006; Lin et al. 1999): **a** space-filling model showing the exterior surface (small subunit in *blue* and the β-sheet domains of the large subunit in *green and red*); **b** interior surface; **c** asymmetric unit of CPMV (small subunit in *blue*; large subunit in *green*); **d** subunit organization, with asymmetric unit outlined in *red*; **e** twofold (*blue oval*), threefold (*blue triangle*), and fivefold (*blue pentagon*) symmetry axes of the icosahedron

example, CPMV particles remain unchanged at 60°C for at least 1 h, and the capsids remain stable through a pH range of 2–12 (Lin and Johnson 2003; Virudachalam and Harrington 1985). In addition to the small percentage of capsids devoid of RNA produced during infection, it is possible to make empty CPMV capsids by hydrolyzing the RNA (Ochoa et al. 2006). Methods for covalent attachments to the interior and exterior surfaces of the CPMV are therefore of use in imparting desired functions to this robust starting platform.

Traditional Bioconjugation Strategies

The chemical techniques brought to bear on viruses and virus-like particles were initially those used routinely for protein derivatization, shown in Fig. 4 (Hermanson 1996; Wong 1991): acylation of the amino groups of lysine side chains and the N-terminus, alkylation of the sulfhydryl group of cysteine, and, to a more limited extent, activation of carboxylic acid residues and coupling with added amines. While these reactions remain the most widely used, the issue of positional selectivity (for example, which lysine(s) of many available lysine choices will be addressed?) joins normal considerations of chemoselectivity (can one address lysine selectively

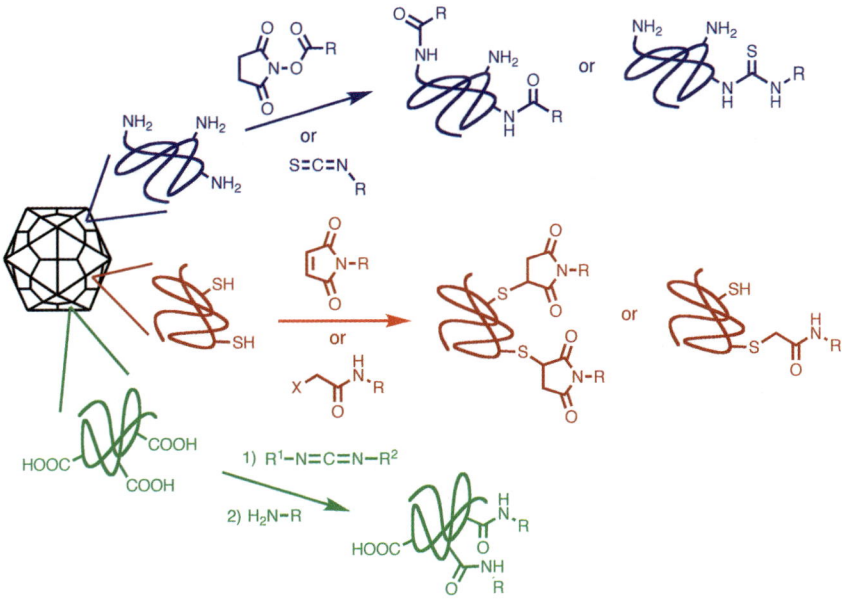

Fig. 4 Traditional bioconjugation methods used for covalent modification of virus particles: (*blue*) acylation of amino groups, usually with *N*-hydroxysuccinimide esters or isothiocyanates; (*red*) alkylation of thiol groups, usually with maleimides or bromo/iodo acetamides; (*black*) activation and capture of carboxylic acid groups using carbodiimides (usually 1-ethyl-3-(3-dimethyll aminopropyl)carbodiimide hydrochloride, *EDC*) and amines

in the presence of other nucleophilic amino acid residues?) and yield when operating on a polyvalent scaffold such as a virion. It therefore was of interest to re-examine the standard reagents in the context of virus reactivity, and CPMV was the first particle used.

Early investigations showed that up to approximately 240 dye molecules could be attached under forcing conditions (Wang et al. 2002a, 2002b), covering most of the surface-exposed lysine side chains (Fig. 5), and confirming earlier indications that the CPMV particle is a relatively static structure that does not expose hidden residues to solvent, in contrast to other particles such as flock house virus (Bothner et al. 1999; Broo et al. 2001). Most interestingly, lysine 38 of the CPMV small subunit was found to be unique in its ability to react with relatively mild isothiocyanate electrophiles due to depressed protic basicity, leaving more of the free amine available in aqueous solution for reaction (Wang et al. 2002a). However, all of the

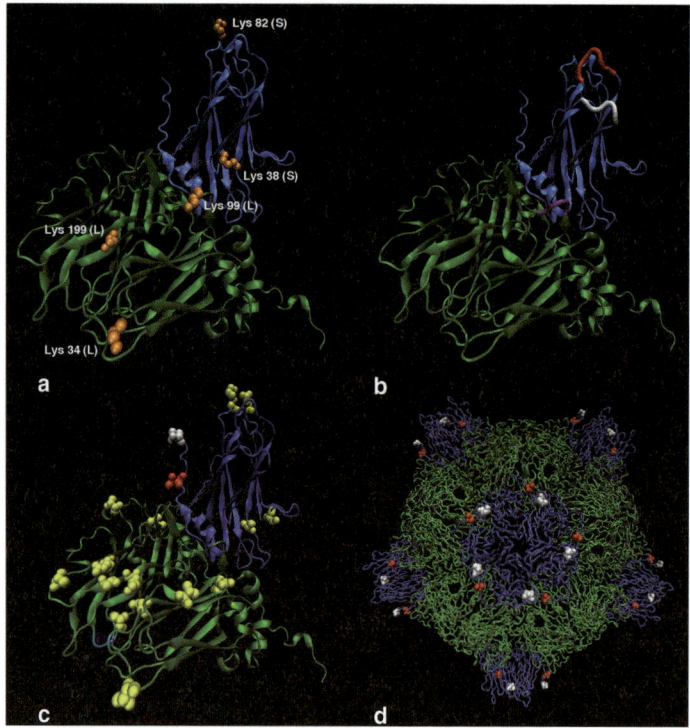

Fig. 5 **a** The CPMV asymmetric unit (small subunit in *blue*; large subunit in *green*) with side chains of the five surface-exposed lysine residues rendered in *orange*. **b** Surface-exposed loops that allow insertion of amino acids into the CPMV capsid structure: BC (*red*), CÎC (*white*) and EF (*purple*). **c** Sites of attempted cysteine point mutations (*yellow*) other than in the loops highlighted in **b**; positions of successful T184C and L189C replacements in *red* and *white*, respectively. **d** View down the fivefold symmetry axis of the CPMV capsid, with space-filling representations of the T184C site in *red* and L189C site in *white*. Note that L189C places the cysteine residue higher (and therefore more exposed) on the capsid protrusion surrounding the fivefold axis

surface lysine residues were shown to react with more potent electrophiles such as NHS esters (Chatterji et al. 2004a). This distributed reactivity could be absolutely controlled only by the construction of mutant (chimeric) particles in which all but one of the surface lysines were changed to arginines. The selective reactivity of this chimera was demonstrated by attachment and visualization of Nanogold. While these studies were successful in using the native lysine for conjugation, little insight was gained into what makes lysine 38 more reactive than others. Our own attempts to examine this question by making changes in surrounding amino acids were frustrated by an inability to express the necessary chimeras in plants, and CPMV is not commonly (Shanks and Lomonossoff 2000; Liu and Lomonossoff 2002) expressed as virus-like particles in any other host.

The x-ray crystal structure of CPMV shows no cysteine residues accessible to solvent on the exterior surface, and interior surface cysteines are either tied up in disulfide linkages or are sterically encumbered by encapsulated RNA. The solution-phase chemistry of the particle proved to be consistent with this picture: CPMV is much less reactive with mild alkylating agents such as bromoacetamides than other proteins having exposed cysteine thiols. This provided an opportunity to introduce reactive cysteines in chimeric structures, and previous work on CPMV provided an excellent guide to this enterprise. Oligopeptide sequences have been inserted into the surface loops of CPMV, to make chimeras primarily for purposes of antibody generation and/or the inhibition of cell surface interactions (Canady et al. 1996; Douglas et al. 1995, 2002; Douglas and Young 1998, 1999, 2005; Douglas 2003; Klem et al. 2005a, 2005b; Radloff et al. 2005; Juhl et al. 2006; Tseng et al. 2006; Niu et al. 1991; Lomonossoff and Johnson 1991; Stauffacher et al. 1987; Bruening and Agrawal 1967; Dessens and Lomonossoff 1993; Lomonossoff 1996; Lin et al. 1996; Lin and Johnson 2003; Virudachalam and Harrington 1985; Ochoa et al. 2006). Three CPMV surface loops, designated BC, CÎC″, and EF, are amenable to peptide insertion (Fig. 5). It has been reported by others, and we have confirmed that the inserted sequences should be shorter then 40 amino acids and not contain repeats in the genetic sequences (which can be edited or duplicated by host recombination pathways).

Our first attempts at cysteine insertion provided highly reactive particles that suffered from concomitant tendencies to aggregate in the absence of high concentrations of reducing agents by formation of interparticle disulfides (Wang et al. 2002b, 2002c). Condensation of these particles with maleimides proceeded in high yield, and labeling with gold clusters followed by cryoelectron microscopy showed that covalent attachments were made at the site of mutation (Wang et al. 2002b, 2002c). Subsequent studies by us, not yet published, have shown that Lys38, previously identified as the most reactive amino group to acylating agents, also competes effectively with inserted cysteines for maleimides, in contrast to conventional wisdom. Such crossreactivity must be kept in mind when the position-selective addressing of polyvalent scaffolds such as virus particles is desired.

Our search for cysteine-insert chimeras that are both reactive toward alkylating agents yet resistant to disulfide-mediated aggregation highlights one of the strengths of bionanoparticles as platforms as well as one of the limitations of CPMV in particular. The strength derives from the power of molecular biology to generate candidate structures in a search for function. More than 20 mutants of CPMV were tested in this case, having cysteines introduced as point mutations over much of the surface topology, with much less effort required to make the particles than to test them. (By comparison, the chemical synthesis of large dendrimers in an analogous exercise would be a Herculean task.) CPMV's weakness from the point of view of chemical exploitation is that it can be expressed on a large scale only in cowpea plants. While convenient from a storage and processing point of view, approximately 2 months of plant growth and virus propagation are therefore required to obtain useful quantities of any new mutant. Even when 20 new particles are produced in parallel in this way, this is too long to wait in many situations. We and others have therefore turned to systems that can be expressed in bacterial cell culture, but CPMV remains quite useful.

Of the chimeras surveyed, two new particles bearing point mutations in the small subunit (T184C and L189C) were found to have improved properties of reactivity and resistance to oxidative aggregation. Both particles resist aggregation in the absence of reducing agent and thereby retain their reactivity indefinitely when stored at moderate concentrations. However, complete cysteine alkylation by maleimides is not possible, even with these particles, before K38 begins to compete. To achieve completely selective attachment on the cysteine residues in T184C and L189C, one must either mutate lysine 38 to arginine or label this residue prior to beginning the maleimide conjugation reaction.

By utilizing the native or the mutationally inserted resides and the standard coupling technologies shown in Fig. 4, a wide variety of materials have been displayed on the surface of CPMV (Table 2). In general, when precise control of the spatial organization of the attached groups is not required and millimolar concentrations of the coupled reagents are available, these methods work well. The yields of recovered particles is generally in the 30%–70% range, although reporting and characterization criteria are not yet completely standardized. We favor the use of both solution-phase (sucrose or cesium chloride gradient ultracentrifugation, size-exclusion and anion-exchange chromatography) and solid-phase (electron microscopy, x-ray crystallography when feasible) methods of characterization of whole particles. Native gel electrophoresis has recently been shown to be useful as well (Steinmetz et al. 2007). The denatured component protein should always be characterized by gel electrophoresis and frequently by protease digestion and mass spectrometry to assure accurate measurement of the number and positions of covalent labels installed.

It is appropriate here to acknowledge the special contributions of Professors George P. Lomonossoff of the John Innes Centre and John E. Johnson of The Scripps Research Institute, and their co-workers. The Lomonossoff laboratory was originally responsible for the genetic characterization (Lomonossoff and Shanks 1983; Zabel et al. 1984) and manipulation of CPMV, the latter by construction of the infectious plasmids (Dessens and Lomonossoff 1993) and associated techniques used by all subsequent workers to develop CPMV mutants. Johnson and co-workers reported the detailed x-ray structural characterization of CPMV (Lin et al. 1999;

Table 2 Modifications of cowpea mosaic virus by the standard bioconjugation methods shown in Fig. 4

Attached species	Residue(s) addressed	Method	Reference
Small molecules	Lysines	NHS ester, isothiocyanates	Wang et al. 2002a; Chatterji et al. 2004; Russell et al. 2005; Steinmetz et al. 2006, 2007
Small molecules	Cysteines	Maleimides, bromoacetamides	Wang et al. 2002c; Sapsford et al. 2006; Soto et al. 2006
Redox-active molecules	Lysines	NHS esters	Steinmetz et al. 2005
Redox-active molecules	Aspartic and glutamic acids	EDC/amines	Steinmetz et al. 2006
Carbohydrates	Lysines, cysteines	Isothiocyanates, bromoacetamides	Raja et al. 2003b
Polyethylene oxide	Lysines	NHS ester	Raja et al. 2003a
Oligonucleotides	Lysines, cysteines	NHS esters, maleimides	Strable et al. 2004
Small proteins (T4 lysozyme, domains of internalin B and herstatin)	Lysines, cysteines	NHS esters, maleimides	Chatterji et al. 2004
Antibodies	Cysteines	Maleimides	Sapsford et al. 2006
Quantum dots	Lysines	NHS ester	Medintz et al. 2005; Blum et al. 2006; Portney et al. 2005
Gold nanoparticles	Cysteines	Au-thiol interaction; maleimides	Wang et al. 2002b; Blum et al. 2004, 2005; Soto et al. 2004
Carbon nanotubes	Lysines	NHS esters	Portney et al. 2005
Nanopatterned surfaces	Cysteines	Maleimides	Cheung et al. 2003; Smith et al. 2003
Solid supports	Lysines – biotin	Biotin-avidin	Steinmetz et al. 2007

Stauffacher et al. 1987), as well as many examples of manipulated CPMV particles for a variety of applications. Both investigators remain very active in this area. For example, in addition to those contributions cited elsewhere in this chapter, Johnson and co-workers have developed useful histidine-tagged versions of CPMV (Chatterji et al. 2005; Cheung et al. 2006) and have used CPMV crystals as templates for materials synthesis (Falkner et al. 2005). Both laboratories have collaborated on the development of genetic inserts that make CPMV a highly immunogenic, and therefore effective, vaccine (Usha et al. 1993; Porta et al. 1994; Dalsgaard et al. 1997; Porta and Lomonossoff 1998; Lomonossoff and Hamilton 1999; Liu et al. 2005), and are also well represented in citations of chemical manipulations in Table 2 and elsewhere. The value of their own contributions has been matched by their generous attitudes in sharing expertise and material with us and many others, thereby furthering the development of CPMV and other particles in an extraordinary range of areas.

Other virus and virus-like particles have been modified in similar ways, although none have crossed boundaries between laboratories to the extent that CPMV has. A brief description of some of the highlights of each follows.

- *Cowpea chlorotic mottle virus* (CCMV) is composed of 180 copies of a single protein subunit. Its chemical reactivity has been explored by Douglas, Young, and co-workers (Gllitzer et al. 2002), in addition to a large and exciting body of work on its use as a nanocapsule for inorganic and magnetic materials synthesis (Suci et al. 2006; Liepold et al. 2005). Unlike CPMV, CCMV can be broken apart into subunits and reassembled into its capsid form. By labeling two populations of CCMV particles with different reagents, disassembling, mixing the two populations, and reassembling, it was possible to create capsids with mixed labels (Gllitzer et al. 2006)
- *Tobacco mosaic virus* (TMV), certainly the longest appreciated virus from a chemical point of view (Crick and Watson 1956), has a helical, rather than icosahedral, arrangement of subunits. Francis and co-workers have recently labeled the interior of TMV, which is lined with glutamic and aspartic acid residues, with a variety of substrates including biotin, chromophores, and crown ethers using carbodiimide coupling reactions (Schlick et al. 2005). The replacement of serine with cysteine residues on the exterior surface allowed for the conjugation of light harvesting dyes to TMV proteins, which were then induced to self-assemble into functional virus-like rods (Miller et al. 2007).
- *Turnip yellow mosaic virus* (TYMV) has been recently introduced to chemical synthesis by the Wang laboratory, which has employed standard NHS ester and carbodiimide/imine coupling to natural and chimeric amino and carboxylic acid containing residues, respectively (Barnhill et al. 2007).
- *Nudaurelia capensis ω virus* (NωV) has also been probed by standard lysine acylation and thiol alkylation reactions (Taylor et al. 2003). This insect virus undergoes a massive conformational change upon proteolytic maturation from a 480-nm diameter procapsid to a 410-nm diameter virion (Taylor et al. 2002). Its response to both chemical reagents (Taylor et al. 2003) and proteolytic enzymes (Bothner et al. 2005) was found to be dramatically affected by this structural rearrangement, with the compact mature particle being much less reactive because it both exposes fewer reactive side chains on average and because it is dynamically less flexible. As is the case with many highly cooperative systems, however, the contributing factors are likely more complicated than this (Bothner et al. 2005).
- Ferritins and heat shock proteins (HSPs) are spherical protein assemblies that are typically smaller than viruses and have fewer subunits, those used for chemical purposes being approximately 12 nm in diameter and composed of 24 identical protein building blocks. They tend to be more chemically stable than viruses and virus-like particles, and can be expressed and purified in quantity. (Several varieties of ferritin are commercially available at modest cost.) The Douglas and

Young laboratories have taken the lead in using these particles for chemical purposes. In an early and spectacular demonstration of the suitability of ferritin to the chemist's bench, exterior carboxylic acids were activated and coupled to long-chain alkylamines to make stable particles that are freely soluble in organic solvents such as dichloromethane. The chemistry of HSP is similarly robust (Flenniken et al. 2003). Ferritin and an HSP have been expressed with tumor-targeting peptides and illuminated by attachment of dyes to aid in tissue imaging (Flenniken et al. 2006), and polymer-modified ferritin has been made for materials self-assembly (Lin et al. 2005).

Tyrosine-Selective Bioconjugation Strategies

In addition to lysine amines and cysteine thiols, the aromatic groups of tyrosine (Hooker et al. 2004; Tilley and Francis 2006; Antos and Francis 2006) and tryptophan (Antos and Francis 2004) have reactivity patterns distinct from the other amino acids, and therefore are attractive targets for bioconjugation. Tyrosines on virus particles have been exploited in three ways, as shown in Fig. 6. The tyrosine phenol is easily oxidized by one electron using peracid or persulfate reagents, mediated by the nickel complex of the gly-gly-his (GGH) tripeptide or by the photochemical action of tris(2,2Î-bipyridyl)ruthenium(II) (Brown and Kodadek 2001; Amini et al. 2005). We exploited this observation by the addition of disulfide trapping agents, giving rise to thioether derivatives of surface-exposed tyrosine residues on CPMV, as well as to intersubunit dityrosine crosslinks within the capsid (Meunier et al. 2004).

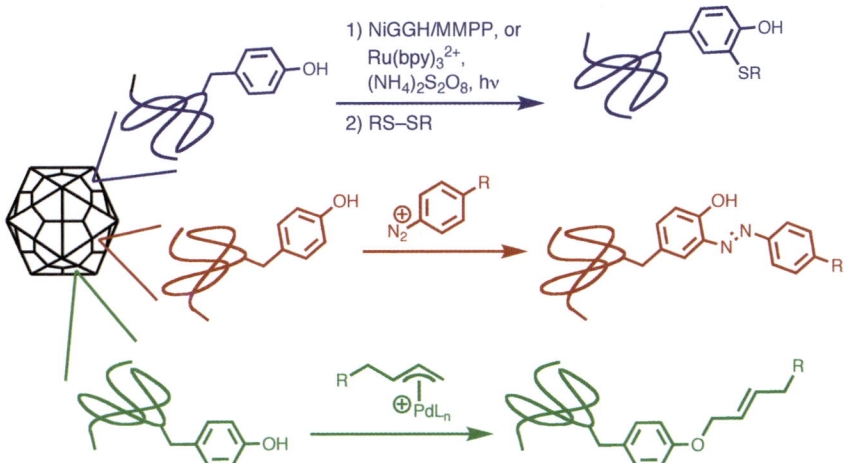

Fig. 6 Methods used for covalent modification of tyrosine residues in virus particles: (*blue*) diazotization; (*red*) one electron oxidation and trapping (*MMPP* magnesium monoperoxyphthalate); (*black*) alkylation by π-allylpalladium complexes derived from allylic acetates and Pd(OAc)$_2$

Francis and co-workers have developed several elegant new methods for bioconjugation and applied them to virus derivatization (Antos and Francis 2006). The reaction of phenols with diazonium salts, long known as a water-friendly reaction in organic synthesis, was exploited to label tyrosine residues of bacteriophage MS2 and TMV (Schlick et al. 2005; Hooker et al. 2004). In the case of MS2, the encapsulated RNA was first hydrolyzed with base, exposing tyrosine residues on the capsid interior. The initial labeling event was followed by a robust and general second connection to the group installed. In addition, the phenolic oxygen of tyrosine can be alkylated by π-allylpalladium complexes formed in situ (Tilley and Francis 2006). Lastly (and not shown in Fig. 6), lysines have been addressed by the Francis group in a new way by reductive alkylation using aldehydes and an iridium transfer hydrogenation catalyst (McFarland and Francis 2005).

Copper(I)-Catalyzed Azide-Alkyne Cycloaddition

Bio-orthogonal reactions – those that involve functional groups that are inert to most biological molecules – have gathered increasing attention for protein conjugation chemistry in general (Agard et al. 2006; Prescher and Bertozzi 2005; van Swieten et al. 2005). By their nature, these processes eliminate potential problems of crossreactivity of electrophilic reagents with biochemical nucleophiles. In the class of bio-orthogonal reagents, (azides + alkynes), (azides + phosphines) (Kiick et al. 2002; Saxon and Bertozzi 2000; Saxon et al. 2002; Mahal et al. 1997), and (aldehydes + hydrazines, hydrazones, or amino ethers) are the most successful. To date, the first pair has been the most widely used with viruses, employing Cu^I or ring strain to accelerate the [3+2] cycloaddition reaction between them, as shown in Fig. 7.

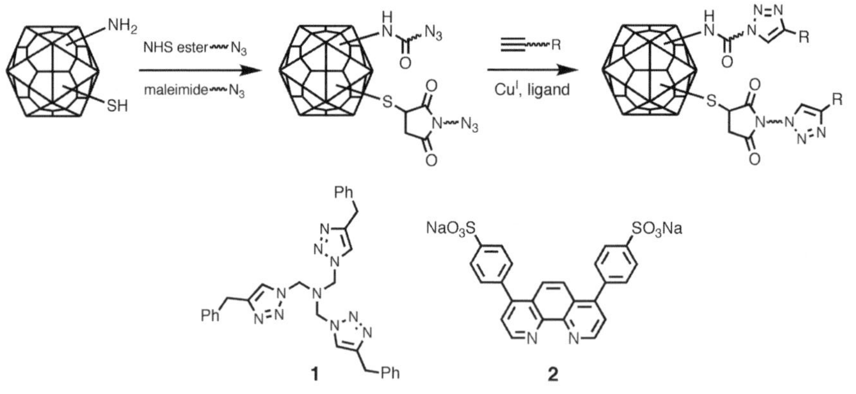

Fig. 7 Installation of azides by standard bioconjugation techniques, followed by coupling with alkynes in the presence of a Cu^I complex involving ligands **1** or **2**. Note that alkynes can be attached to the particle and then coupled with azides in the same way

In most of the applications so far, azides and alkynes are introduced to the virus scaffold by one of the methods described above. The positional control is therefore the same as before, which is to say that it varies widely with the particular scaffold and reaction employed. The subsequent azide-alkyne cycloaddition step, however, is perfectly selective and very rapid. In order for this reaction to be fully compatible with biomolecules in vitro if not yet in vivo (Agard et al. 2006), a copper binding ligand is required to accelerate the reaction, minimize the oxidation of copper from the +1 to +2 states, and prevent the metal from inducing protein aggregation or degradation (Wang et al. 2003) Both the tris(triazolylmethyl)amine **1** (Wang et al. 2003; Chan et al. 2004) and sulfonated bathophenanthroline ligand **2** (Lewis et al. 2004; Sen Gupta et al. 2005a) have been used for the conjugation of a variety of molecules to viruses. Neither ligand is ideal: **1** supports efficient catalysis of the reaction but has marginal water solubility, while **2** is fully water soluble and makes a faster catalyst, but makes the catalytic system much more sensitive to oxygen and therefore must be used in an anaerobic environment. In our hands, **2** allows at least a tenfold reduction in the amount of coupling partner needed to fully address the virus-azide or -alkyne reactant compared to other bioconjugation methods. This improved efficiency has expanded the array of substances that can be attached to viral scaffolds. CPMV has been addressed in this manner with small molecules such as fluorescent dyes (Meunier et al. 2004; Wang et al. 2003; Sen Gupta et al. 2005a), gadolinium complexes (Prasuhn et al. 2007), sugars, polymers (Sen Gupta et al. 2005b) and even the 80-kDa protein transferrin (Sen Gupta et al. 2005a).

In order to position attached structures with precision on virus surfaces, we have recently incorporated unnatural amino acids containing azide or alkyne side chains into capsid proteins under genetic control. Utilizing the auxotroph technology developed by Tirrell and co-workers (Kiick et al. 2000, 2001, 2002; Kiick and Tirrell 2000), sense codon reassignment was used to incorporate azidohomoalanine in place of methionine in both the hepatitis B virus (HBV) particle and the bacteriophage Qβ capsids expressed in *Escherichia coli* (E. Strable et al., unpublished data). Tight control over protein expression resulted in high yields of specifically labeled material, and subsequent azide-alkyne cycloaddition to these positions occurs smoothly.

Conclusions

The chemical manipulation of virus-like particles is always done for the purpose of bringing new properties to these matchless scaffolds. As with all branches of chemical synthesis, familiar bioconjugation reactions were used first and continue to be used most often. New techniques fill much-needed capabilities of chemoselectivity, rate, and positional selectivity for certain applications. The combination of biological and chemical capabilities, such as the introduction of unusually reactive natural residues such as cysteine or unnatural amino acids containing orthogonally reactive groups, takes maximal advantage virus-like particles as bridges between the worlds of biology and chemistry. In this unique way, molecular biology contributes to chemical synthesis on the chemist's scale, to the benefit of drug discovery, drug delivery, materials science, nanotechnology, and other pursuits.

References

Agard NJ, Baskin JM, Prescher JA, Lo A, Bertozzi CR (2006) A comparative study of bioorthogonal reactions with azides. ACS Chem Biol 1:644–648

Amini F, Denison C, Lin H-J, Kuo L, Kodadek T (2003) Using oxidative crosslinking and proximity labeling to quantitatively characterize protein-protein and protein-peptide complexes. Chem Biol 10:1115–1127

Antos JM, Francis MB (2004) Selective tryptophan modification with rhodium carbenoids in aqueous solution. J Am Chem Soc 126:10256

Antos JM, Francis MB (2006) Transition metal catalyzed methods for site-selective protein modification. Curr Opin Chem Biol 10:253–262

Ashcroft AE, Lago H, Macedo JM et al. (2005) Engineering thermal stability in RNA phage capsids via disulphide bonds. J Nanosci Nanotech 5:2034–20418

Barnhill H, Reuther R, Ferguson PL, Dreher TW, Wang Q (2007) Turnip yellow mosaic virus as a chemoaddressable bionanoparticle. Bioconj Chem 18:852–859

Blum AS et al. (2004) Cowpea mosaic virus as a scaffold for 3-D patterning of gold nanoparticles. Nano Lett 4:867–870

Blum AS et al. (2005) An engineered virus as a scaffold for three-dimensional self-assembly on the nanoscale. Small 1:702–706

Blum AS et al. (2006) Templated self-assembly of quantum dots from aqueous solution using protein scaffolds. Nanotechnology 17:5073–5079

Bothner B, Schneemann A, Marshall D et al. (1999) Crystallographically identical virus capsids display different properties in solution. Nat Struct Biol 6:114–116

Bothner B, Taylor D, Jun B et al. (2005) Maturation of a tetravirus capsid alters the dynamic properties and creates a metastable complex. Virology 334:17–27

Broo K, Wei J, Marshall D et al. (2001) Viral capsid mobility: a dynamic conduit for inactivation. Proc Nat Acad Sci U S A 98:2274–2277

Brown KC, Kodadek T (2001) Protein cross-linking mediated by metal ion complexes. Metal Ions Biol Sys 38:351–384

Bruening GE, Agrawal HO (1967) Infectivity of a mixture of cowpea mosaic virus ribonucleoprotein components. Virology 32:306–320

Canady MA, Larson SB, Day J, McPherson A (1996) Crystal structure of turnip yellow mosaic virus. Nat Struct Biol 3:771–781

Chan TR, Hilgraf R, Sharpless KB, Folkin VV (2004) Polytriazoles as copper(I)-stabilizing ligands in catalysis. Org Lett 6:2853

Chatterji A, Ochoa WF, Paine F et al. (2004a) New addresses on an addressable virus nanoblock uniquely reactive lys residues on cowpea mosaic virus. Chem Biol 11:855–863

Chatterji A, Ochoa W, Shamieh L et al. (2004b) Chemical conjugation of heterologous proteins on the surface of cowpea mosaic virus. Bioconj Chem 15:807–813

Chatterji A, Ochoa WF, Ueno T, Lin T, Johnson JE (2005) A virus-based nanoblock with tunable electrostatic properties. Nano Lett 5:597–602

Cheung CL, Camarero JA, Woods BW et al. (2003) Fabrication of assembled virus nanostructures on templates of chemoselective linkers formed by scanning probe nanolithography. J Am Chem Soc 125:6848–6849

Cheung CL, Chung SW, Chatterji A et al. (2006) Physical controls on directed virus assembly at nanoscale chemical templates. J Am Chem Soc 128:10801–10807

Chillon M, Lee JH, Fasbender A, Welsh MJ (1998) Adenovirus complexed with polyethylene glycol and cationic lipid is shielded from neutralizing antibodies in vitro. Gene Ther 5:995–1002

Crick FHC, Watson JD (1956) Structure of small viruses. Nature 177:473–475

Dalsgaard K, Uttenthal A, Jones TD et al. (1997) Plant-derived vaccine protects target animals against a viral disease. Nat Biotech 15:248–252

Dessens JT, Lomonossoff GP (1993) Cauliflower mosaic 35S promoter-controlled DNA copies of cowpea mosaic virus RNAs are infectious on plants. J Gen Virol 74:889–892

Douglas T (2003) Materials science. A bright bio-inspired future. Science 299:1192–1193

Douglas T, Young M (1998) Host-guest encapsulation of materials by assembled virus protein cages. Nature 393:152–155

Douglas T, Young M (1999) Virus particles as templates for materials synthesis. Adv Mater 11:679–681

Douglas T, Young M (2006) Viruses: making friends with old foes. Science 312:873–875

Douglas T, Dickson DPE, Betteridge S et al. (1995) Synthesis and structure of an iron(III) sulfide-ferritin bioinorganic nanocomposite. Science 269:54–57

Douglas T, Strable E, Willits D et al. (2002) Protein engineering of a viral cage for constrained nanomaterials synthesis. Adv Mater 14:415–418

Falkner JC, Turner ME, Bosworth JK et al. (2005) Virus crystals as nanocomposite scaffolds. J Am Chem Soc 127:5274–5275

Flenniken ML, Willits DA, Brumfield S, Young MJ, Douglas T (2003) The small heat shock protein cage from *Methanococcus jannaschii* is a versatile nanoscale platform for genetic and chemical modification. Nano Lett 3:1573–1576

Flenniken ML, Willits DA, Harmsen AL et al. (2006) Melanoma and lymphocyte cell-specific targeting incorporated into a heat shock protein cage architecture. Chem Biol 13:161–170

Gillitzer E, Willits D, Young M, Douglas T (2002) Chemical modification of a viral cage for multivalent presentation. Chem Commun 2390–2391

Gillitzer E, Suci PA, Young Mark J, Douglas ES (2006) Controlled ligand display on a symmetrical protein-cage architecture through mixed assembly. Small 2:962–966

Golmohammadi R, Valegard K, Fridborg K, Liljas L (1993) The redefined structure of bacteriophage MS2 at 2.8A resolution. J Mol Biol 234:620–639

Golmohammadi R, Fridborg K, Bundule M, Liljas L (1996) The crystal structure of bacteriophage Q beta at 3.5A resolution. Structure 4:543–554

Granier T, Gallois B, Dautant A, Estaintot BLD, Precigoux G (1997) Comparison of the structures of the cubic and tetragonal forms of horse-spleen apoferritin. Acta Cryst D 53:580–587

Hermanson GT (1991) Bioconjugate techniques. Academic Press, San Diego

Hooker JM, Kovacs EW, Francis MB (2004) Interior surface modification of bacteriophage MS2. J Am Chem Soc 126:3718–3719

Juhl SB, Chan EP, Ha YH et al. (2006) Assembly of Wiseana iridovirus: viruses for colloidal photonic crystals. Adv Func Mater 16:1086–1094

Kiick KL, Tirrell DA (2000) Protein engineering by in vivo incorporation of non-natural amino acids: control of incorporation of methioonine analogues by methionyl tRNA synthetase. Tetrahedron 56:9487–9493

Kiick KL, Van Hest JCM, Tirrell DA (2000) Expanding the scope of protein biosynthesis by altering the methionyl-tRNA synthetase activity of a bacterial expression host. Angew Chem Int Ed Engl 39:2148–2152

Kiick KL, Weberskirch R, Tirrell DA (2001) Identification of an expanded set of translationally active methionine analogues in *Escherichia coli*. FEBS Lett 502:25–30

Kiick KL, Saxon E, Tirrell DA, Bertozzi CR (2002) Incorporation of azides into recombinant proteins for chemoselective modification by the Staudinger ligation. Proc Natl Acad Sci U S A 99:19–24

Kim KK, Kim R, Kim SH (1998) Crystal structure of a small heat-shock protein. Nature 394:595–599

Klem MT, Young M, Douglas T (2005a) Biomimetic magnetic nanoparticle. Mater Today 8:28–37

Klem MT, Willits D, Solis DJ et al. (2005b) Bio-inspired synthesis of protein-encapsulated CoPt nanoparticles. Adv Func Mater 15:1489–1494

Kuhn RJ, Zhang W, Rossman MG et al. (2002) Structure of dengue virus: implications for flavivirus organization, maturation, and fusion. Cell 108:717–725

Lewis WG, Magallon FG, Fokin VV, Finn MG (2004) Discovery and characterization of catalysts for azide-alkyne cycloaddition by fluorescence quenching. J Am Chem Soc 126:9152–9153

Liepold LO, Revis J, Allen M et al. (2005) Structural transitions in cowpea chlorotic mottle virus (CCMV). Phys Biol 2:S166–S172

Lin T, Johnson JE (2003) Structure of picorna-like plant viruses: implications and applications. Adv Virus Res 62:167–239

Lin T, Porta C, Lomonossoff G, Johnson JE (1996) Structure-based design of peptide presentation on a viral surface: the crystal structure of a plant/animal virus chimera at 2.8 .ANG resolution. Fold Des 1:179–187

Lin T, Chen Z, Usha R et al. (1999) The refined crystal structure of cowpea mosaic virus at 2.8 Å resolution. Virology 265:20–34

Lin Y, Böker A, He J et al. (2005) Self-directed self-assembly of nanoparticle/copolymer mixtures. Nature 434:55–59

Liu L, Lomonossoff GP (2002) Agroinfection as a rapid method for propagating Cowpea mosaic virus-based constructs. J Virol Methods 105:343–348

Liu L, Cañizares MC, Monger W et al. (2005) Cowpea mosaic virus-based systems for the production of antigens and antibodies in plants. Vaccine 23:1788–1792

Lomonossoff GP (1996) Modified plant viruses as vectors of heterologous peptides and use as animal vaccines. In: PCT Int. Appl. Axis Genetics Ltd., UK

Lomonossoff GP, Hamilton WDO (1999) Cowpea mosaic virus-based vaccines. Curr Topics Microbiol Immun 240:177–189

Lomonossoff GP, Johnson JE (1991) The synthesis and structure of comovirus capsids. Prog Biophys Mol Biol 55:107–137

Lomonossoff GP, Shanks M (1983) The nucleotide sequence of cowpea mosaic virus B RNA. EMBO J 2:2253–2258

Mahal LK, Yarema KJ, Bertozzi CR (1997) Engineering chemical reactivity on cell surfaces through oligosaccharide biosynthesis. Science 276:1125–1128

Marlow SA, Delgado C, Neale D, Francis GE (1999) ViraMASC: a biologically optimized pegylation technology to target adenovirus to tumors. Proc Int Symp Controlled Release Bioact Mater 26:555–556

Martin CS et al. (2001) Combined EM/X-ray imagining yields a quasi-atomic model of the adenovirus-related bacteriophage PRD1 and shows key capsid and membrane interactions. Structure 9:917–930

McFarland JM, Francis MB (2005) Reductive alkylation of proteins using iridium catalyzed transfer hydrogenation. J Am Chem Soc 127:13490–13491

Medintz IL, Sapsford KE, Konnert JH et al. (2005) Decoration of discretely immobilized cowpea mosaic virus with luminescent quantum dots. Langmuir 21:5501–5510

Meldrum FC, Wade VJ, Nimmo DL, Heywood BR, Mann S (1991) Synthesis of inorganic nanophase materials in supramolecular protein cages. Nature 349:684–687

Meunier S, Strable E, Finn MG (2004) Crosslinking of and coupling to viral capsid proteins by tyrosine oxidation. Chem Biol 11:319–326

Miller RA, Preseley AD, Francis MB (2007) Self-assembling light harvesting systems from synthetically modified tobacco mosaic virus. J Am Chem Soc 129:3104–3109

Munshi S, Liljas L, Johnson JE (1998) Structure determination of *Nudaurelia capensis* omega virus. Acta Cryst D 54:1295–1305

Nakagawa A, Miyazaki N, Taka J et al. (2003) The atomic structure of rice dwarf virus reveals the self-assembly mechanism of component proteins. Structure 11:1227–1238

Namba K, Caspar DLD, Stubbs G (1985) Computer graphics representation of levels of organization in tobacco mosaic virus structure. Science 227:773–776

Niu Z, Bruckman M, Kotakadi VS et al. (2006) Study and characterization of tobacco mosaic virus head-to-tail assembly assisted by aniline polymerization. Chem Commun (Camb) 3019–3021

Ochoa WF, Chatterji A, Lin T, Johnson JE (2006) Generation and structural analysis of reactive empty particles derived from an icosahedral virus. Chem Biol 13:771–778

O'Riordan CR, Lachapelle A, Delgado C et al. (1999) PEGylation of adenovirus with retention of infectivity and protection from neutralizing antibody in vitro and in vivo. Hum Gene Ther 10:1349–1358

Paillard F (1999) Dressing up adenoviruses to modify their tropism. Hum Gene Ther 10:2575–2576

Porta C, Lomonossoff GP (1998) Scope for using plant viruses to present epitopes from animal pathogens. Rev Med Virol 8:25–41

Porta C, Spall VE, Loveland J et al. (1994) Development of cowpea mosaic virus as a high-yielding system for the presentation of foreign peptides. Virology 202:949–955

Portney NG, Singh K, Chaudhary S et al. (2005) Organic and inorganic nanoparticle hybrids. Langmuir 21:2098–2103

Prasad BV, Hardy ME, Dokland T et al. (1999) X-ray crystallographic structure of the Norwalk virus capsid. Science 286:287–290

Prasuhn J, DE, Yeh RM, Obenaus A, Manchester M, Finn MG (2007) Viral MRI contrast agents: coordination of Gd by native virions and attachment of Gd complexes by azide-alkyne cycloaddition. Chem Commun 1269–1271

Prescher JA, Bertozzi CR (2005) Chemistry in living systems. Nat Chem Biol 1:13–21

Radloff C, Vaia RA, Brunton J, Bouwer GT, Ward VK (2005) Metal nanoshell assembly on a virus bioscaffold. Nano Lett 5:1187–1191

Raja KS, Wang Q, Finn MG (2003a) Icosahedral virus particles as polyvalent carbohydrate display platforms. ChemBioChem 4:1348–1351

Raja KS, Wang Q, Gonzalez MJ et al. (2003b) Hybrid virus-polymer materials. 1. Synthesis and properties of peg-decorated cowpea mosaic virus. Biomacromolecules 4:472–476

Russell JT, Lin Y, Böker A et al. (2005) Self-assembly and cross-linking of bionanoparticles at liquid-liquid interfaces. Angew Chem Int Ed 44:2420–2426

Sapsford KE, Soto CM, Blum AS et al. (2006) A cowpea mosaic virus nanoscaffold for multiplexed antibody conjugation: application as an immunoassay tracer. Biosens Bioelectron 21:1668–1673

Saxon E, Bertozzi Carolyn R (2000) Cell surface engineering by a modified Staudinger reaction. Science 287:2007–2010

Saxon E, Luchansky SJ, Hang SC et al. (2002) Investigating cellular metabolism of synthetic azidosugars with the Staudinger Ligation. J Am Chem Soc 124:14893

Schlick TL, Ding Z, Kovacs EW, Francis MB (2005) Dual-surface modification of the tobacco mosaic virus. J Am Chem Soc 127:3718–3723

Sen Gupta S, Kuzelka J, Singh P et al. (2005a) Accelerated bioorthogonal conjugation: a practical method for the ligation of diverse functional molecules to a polyvalent virus scaffold. Bioconj Chem 16:1572–1579

Sen Gupta S, Raja KS, Kaltgrad E, Strable E, Finn MG (2005b) Virus-glycopolymer conjugates by copper(I) catalysis of atom transfer radical polymerization and azide-alkyne cycloaddition. Chem Commun 4315–4317

Shanks M, Lomonossoff GP (2000) Co-expression of the capsid proteins of cowpea mosaic virus in insect cells leads to the formation of virus-like particles. J Gen Virol 81:3093–3097

Shepherd CM, Borelli IA, Lander G et al. (2006) VIPERdb: a relational database for structural virology. Nucleic Acids Res 34:D386–D389

Smith JC et al. (2003) Nanopatterning the chemospecific immobilization of cowpea mosaic virus capsid. Nano Lett 3:883–886

Soto CM et al. (2004) Separation and recovery of intact gold-virus complex by agarose electrophoresis and electroelution: application to the purification of cowpea mosaic virus and colloidal gold complex. Electrophoresis 25:2901–2906

Soto CM, Blum AS, Vora GJ et al. (2006) Fluorescent signal amplification of carbocyanine dyes using engineered viral nanoparticles. J Am Chem Soc 128:5184–5189

Speir JA, Munshi S, Wang G, Baker TS, Johnson JE (1995) Structures of the native and swollen forms of cowpea chlorotic mottle virus determined by X-ray crystallography and cryo-electron microscopy. Structure 3:63–78

Stauffacher CV et al. (1987) The structure of cowpea mosaic virus at 3.5 Ang. resolution. In: Moras J, Drenth J, Strandberg B, Suck D, Wilson K (eds) Crystallography in molecular biology. Plenum, New York, pp 293–308

Steinmetz NF, Lomonossoff GP, Evans DJ (2005) Decoration of cowpea mosaic virus with multiple redox active organometallic complexes. Small 2:530–533

Steinmetz NF, Calder G, Lomonossoff G, Evans DJ (2006) Plant viral capsids as nanobuilding blocks: construction of arrays on solid supports. Langmuir 22:10032–10037

Steinmetz NF, Lomonossoff GP, Evans DJ (2006) Cowpea mosaic virus for material fabrication: addressable carboxylate groups on a programmable nanoscaffold. Langmuir 22:3488–3490

Steinmetz NF, Evans DJ, Lomonossoff GP (2007) Chemical introduction of reactive thiols into a viral nanoscaffold: a method which avoids virus aggregation. ChemBioChem 8:1131–1136

Steinmetz NF, Evans DJ, Lomonossoff GP (2007) Monitoring aggregation of chemically and genetically engineered thiol-decorated viral nanoparticles. ChemBioChem, in press

Strable E, Johnson JE, Finn MG (2004) Natural nanochemical building blocks: icosahedral virus particles organized by attached oligonucleotides. Nano Lett 4:1385–1389

Suci PA, Klem MT, Arce FT, Douglas T, Young M (2006) Assembly of multilayer films incorporating a viral protein cage architecture. Langmuir 22:8891–8896

Taylor DJ, Krishna NK, Canady MA, Schneemann A, Johnson JE (2002) Large-scale, pH-dependent, quaternary structure changes in an RNA virus capsid are reversible in the absence of subunit autoproteolysis. J Virol 76:9972–9980

Taylor DJ, Wang Q, Bothners B et al. (2003) Correlation of chemical reactivity of Nudaurelia capensis omega virus with a pH-induced conformational change. Chem Commun 2770–2771

Tilley SD, Francis MB (2006) Tyrosine-selective protein alkylation using p-allylpalladium complexes. J Am Chem Soc 128:1080–1081

Tseng RJ, Tsai C, Ma L et al. (2006) Digital memory device based on tobacco mosaic virus conjugated with nanoparticles. Nat Nanotech 1:72–77

Usha R, Rholl JB, Spall VE et al. (1993) Expression of an animal virus antigenic site on the surface of a plant virus particle. Virology 197:366–374

van Swieten PF, Leeuwenburgh MA, Kessler BM, Overkleeft HS (2005) Bioorthogonal organic chemistry in living cells: novel strategies for labeling biomolecules. Org Biomol Chem 3:20–27

Virudachalam R, Harrington MM (1985) Thermal stability of cowpea mosaic virus components: differential scanning calorimetry studies. Virology 146:138–140

Wang Q, Kaltgrad E, Lin T, Johnson JE, Finn MG (2002a) Natural supramolecular building blocks: wild-type cowpea mosaic virus. Chem Biol 9:805–811

Wang Q, Lin T, Tang L, Johnson JE, Finn MG (2002b) Icosahedral virus particles as addressable nanoscale building blocks. Angew Chem Int Ed 41:459–462

Wang Q, Lin T, Johnson JE, Finn MG (2002c) Natural supramolecular building blocks cysteine-added mutants of cowpea mosaic virus. Chem Biol 9:813–819

Wang Q, Chan TR, Hilgraf R et al. (2003) Bioconjugation by copper(I)-catalyzed azide-alkyne [3+2] cycloaddition. J Am Chem Soc 125:3192–3193

Wikoff WR, Duda RL, Hendrix RW, Johnson JE (1999) Crystallographic analysis of the dsDNA bacteriophage HK97 mature empty capsid. Acta Crystallogr D: Biol Crystallogr D55:763–771

Wong SS (1991) Chemistry of protein conjugation and cross-linking. CRC Press, Boca Raton, FL

Wynne SA, Crowther RA, Leslie AG (1999) The crystal structure of the human hepatitis B virus capsid. Mol Cell 3:771–780

Zabel P, Moerman M, Lomonossoff G, Shanks M, Beyreuther K (1984) Cowpea mosaic virus VPg: sequencing of radiochemically modified protein allows mapping of the gene on B RNA. EMBO J 3:1629–1634

Zalipsky S (1995) Chemistry of polyethylene-glycol conjugates with biologically active molecules. Adv Drug Del Rev 16:157–182

Structure-Based Engineering of an Icosahedral Virus for Nanomedicine and Nanotechnology

N.F. Steinmetz, T. Lin, G.P. Lomonossoff, J.E. Johnson (✉)

Contents

Introduction	24
Assembly of Nanoparticles with Designed Antigenicity	26
Design and Generation of CPMV Chimeras	27
Imaging of CPMV Chimeras by X-Ray Crystallography	28
CPMV Chimeras as the Vaccine Candidates	30
CPMV Particles Containing Heterologous RNA as Diagnostic Reagents	33
Chemical Fabrications of CPMV-Based Assemblies	34
Assembly of CPMV–Gold Conjugates by Thiol Chemistry	34
Engineering of Specific Amine Reactivity for Assembly of Supramolecules	37
Electroactive CPMV Complexes	42
Display of Redox-Active Complexes on CPMV	42
Conducting 3D Networks on CPMV	45
Targeted Delivery to Mammalian Cells by the Attachment of Functional Proteins	46
Fabrication of CPMV Arrays	47
3D CPMV Arrays via Crystallization Procedures	49
3D CPMV Arrays via Layer-by-Layer Assembly	50
2D CPMV Arrays	54
Perspective	55

Abstract A quintessential tenet of nanotechnology is the self-assembly of nanometer-sized components into devices. Biological macromolecular systems such as viral particles were found to be suitable building blocks for nanotechnology for several reasons: viral capsids are extremely robust and can be produced in large quantities with ease, the particles self-assemble into monodisperse particles with a high degree of symmetry and polyvalency, they have the propensity to form arrays, and they offer programmability through genetic and chemical engineering. Here, we review the recent advances in engineering the icosahedral plant virus

---

J.E. Johnson
Department of Molecular Biology, The Scripps Research Institute, 10550 North Torrey Pines Road, La Jolla, CA 92037, USA
e-mail: jackj@scripps.edu

Cowpea mosaic virus (CPMV) for applications in nano-medicine and -technology. In the first part, we will discuss how the combined knowledge of the structure of CPMV at atomic resolution and the use of chimeric virus technology led to the generation of CPMV particles with short antigenic peptides for potential use as vaccine candidates. The second part focuses on the chemical addressability of CPMV. Strategies to chemically attach functional molecules at designed positions on the exterior surface of the viral particle are described. Biochemical conjugation methods led to the fabrication of electronically conducting CPMV particles and networks. In addition, functional proteins for targeted delivery to mammalian cells were successfully attached to CPMV. In the third part, we focus on the utilization of CPMV as a building block for the generation of 2D and 3D arrays. Overall, the potential applications of viral nanobuilding blocks are manifold and range from nanoelectronics to biomedical applications.

Abbreviations CPMV: Cowpea mosaic virus; EDC: N-ethyl-N'-(3-dimethylaminopropyl)carbodiimide hydrochloride; EGFR: Epidermal growth factor receptor; FITC: Fluorescein isothiocyanate; NHS: N-hydroxysuccinimide; QCMD: Quartz crystal microbalance with dissipation monitoring

Introduction

A quintessential tenet of nanotechnology is the self-assembly of nanometer-sized components into devices. Recent advances in synthetic chemistry led to the fabrication of a range of interesting molecules and nanoscale components; however, functional connectivity among different components in a predefined pattern is difficult to achieve. Biological macromolecule systems are generally more amenable for self-assembly, not only because of their natural propensity to form arrays, but also because they offer programmability through genetic engineering. They can be utilized either directly as devices or indirectly as templates for patterning other small synthetic or biological molecules. Another benefit of biological macromolecules compared to synthetic nanomaterials is biocompatibility; this is particularly important for biological or medical applications. The ideal properties of a biological system for applications in nanosciences and nanotechnology include high yield, structural definition to atomic resolution coupled with a high degree of chemical stability. We have exploited icosahedral viruses, especially the plant virus, Cowpea mosaic virus (CPMV), for applications in nanomedicine and nanotechnology. Besides having the above-mentioned properties, icosahedral virus particles also possess a high degree of symmetry, leading to polyvalency and the capacity to carry large cargos and extensive surfaces for functional engineering.

CPMV is a picorna-like virus with a genome of two segments of single-stranded, positive sense RNA (Fig. 1). The larger RNA-1 encodes the virus replication machinery and the smaller RNA-2 encodes the two capsid proteins and the viral movement protein. The two RNA molecules are separately encapsidated in isometric

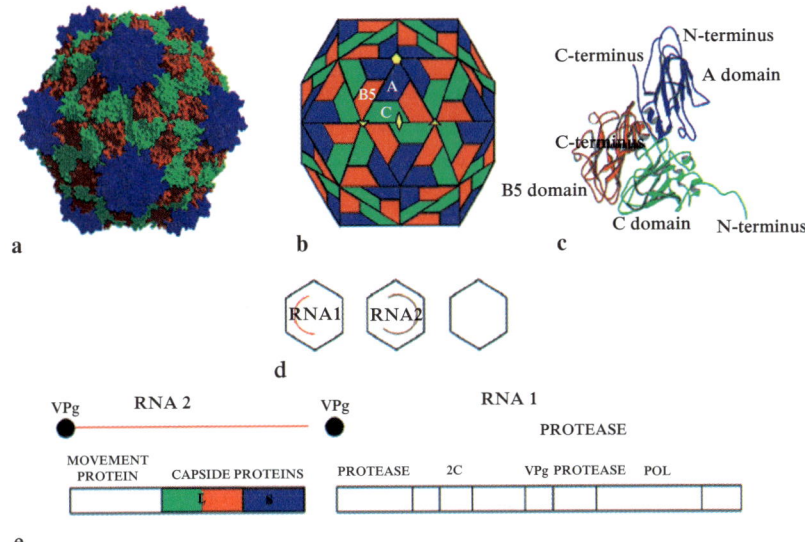

Fig. 1 **a** A space-filling drawing of the CPMV capsid. **b** A schematic presentation of CPMV capsid. The capsid is comprised by two viral proteins, the S and L subunits, which form three β-sandwich domains in the icosahedral asymmetric unit. The S subunit occupies A (*blue*) positions around the fivefold axis; the two domains of L subunit occupy the B5 (*red*) and C (*green*) positions. Sixty copies of the S and L subunits comprised the viral capsid. **c** A ribbon diagram of the three β-barrel domains that comprise the icosahedral asymmetric unit. **d** The two RNA molecules of the virus genome are separately encapsidated and both types of the particles are required for infection. Empty particles are also formed. **e** Two RNA molecules, RNA-1 and RNA-2, comprise the CPMV genome, with RNA-2 encoding S and L capsid proteins

particles, and both types of particles are required for infection. Empty virus particles containing no RNA are also produced during an infection (Lomonossoff and Johnson 1991).

An essential component for the generation of genetically engineered viral particles is the availability of infectious cDNA clones. The first infectious clones of CPMV were based on the generation of RNA transcripts in vitro by *Escherichia coli* or T7 polymerases (Eggen et al. 1989; Holness et al. 1989; Rohll et al. 1993; Vos et al. 1998). A significant improvement was made with the introduction of clones based on the use of the 35S promoter of *Cauliflower mosaic virus* (Dessens and Lomonossoff 1993), which were subsequently adapted for delivery by *Agrobacterium tumefaciens* (Liu and Lomonossoff 2002). As the host RNA polymerase recognizes the promoter to generate viral RNA transcripts in situ, mutated cDNA encoding the CPMV genome can be mechanically introduced to plants for the production of engineered viruses with unparalleled convenience and efficiency (Fig. 2). The yield of CPMV is 1–2 g of virus per kilogram of infected leaves and the virus can be easily purified by a straightforward protocol (Wellink 1998). The virus particles are substantially more stable than a typical macromolecular

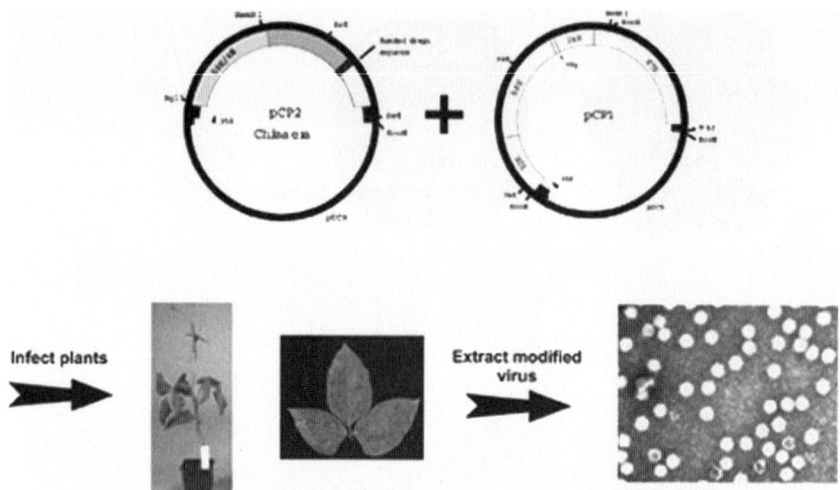

Fig. 2 Scheme for the production of CPMV-based chimeras. Two cDNA infectious clones each encoding one of the viral RNA molecules are under the control of the 35S promoter from *Cauliflower mosaic virus*. Mechanical inoculation of both cDNA onto cowpea plants sets off the viral infection for the production of CPMV. Gram quantities of CPMV can be isolated in laboratory setting. (Johnson et al. 1997, with permission)

assembly. The native virus can tolerate organic solvents such as dimethyl sulfoxide at concentrations up to 50% (v/v) for at least 2 days and maintains the integrity and infectivity for months at room temperature, for days at 37°C, and for hours at 60°C (Wang et al. 2002).

Knowledge of the 3D structure is essential for rational design of nanomaterials. The crystal structure of CPMV was determined and refined to near atomic resolution (Lin et al. 1999). The viral capsid is comprised of two proteins subunits, the small (S) and large (L) subunit. The S subunit is about 23 kD and folds into a jellyroll β-sandwich, while the L subunit of folds into two jellyroll β-sandwiches, with a total mass of 41 kD. Sixty copies of each of the capsid proteins form the virus capsid of 30 nm in a $P=3$ symmetry, with the asymmetric unit consisting of one copy each of an L and S subunit (Fig. 1).

Assembly of Nanoparticles with Designed Antigenicity

With a well-defined structure and readily programmable capsid, CPMV particles are exceptional nanobuilding blocks for the fabrication of nanoassemblies. The simplest device can be made with a single type of an exogenous component assembled on a CPMV scaffold. CPMV particles decorated with short antigenic peptides are such nanoassemblies which were exploited for the generation of vaccines. Since the constructs are made through genetic engineering, no special fabrication is required other than propagation and purification of the chimeric viruses as carried out for any other CPMV mutant. There are several advantages in using the CPMV system for vaccine production: the antigenic potency of the peptide is enhanced by the polyvalent presentation due to the high symmetry of the virus particles; the ease of production and high yield make vaccines produced in this way very affordable; the stability of the CPMV carrier allows long-term storage and convenience in transportation and distribution.

Design and Generation of CPMV Chimeras

There are several prominent and permissive locations on the CPMV surface for the presentation of foreign peptides. The βB-βC and βC'-βC" loops of the S subunit are the equivalents of NIm-IA antigenic site and the βE-βF loop of the L subunit is the location of NIm-II site in human rhinoviruses. A chimeric virus technology was developed in which exogenous peptides were inserted by cassette mutagenesis and the progeny virus particles carrying the antigenic peptides were obtained by infecting plants with modified versions of the infectious clones (Fig. 3) (Porta et al. 1994; Taylor et al. 2000).

The usefulness and versatility of the system was demonstrated by the investigation of the antigenic assemblies carrying a peptide of 14 residues (KDATGIDNHREAKL) corresponding to the NIm1A epitope from VP1 of human rhinovirus (HRV) 14. The initial construct was made by the insertion of the peptide between the Ala122 and Pro123 of the βB-βC loop of CPMV S subunit to produce a chimera called CPMV/HRV-II (Fig. 3) (Porta et al. 1994). Additional constructs were made by moving the insertion site one residue to the left (between Pro121 and Ala122) to produce CPMV/HRV-L1 (Taylor et al. 2000). Similarly, by moving the insertion site to one, two and three residues to the right, chimeras CPMV/HRV-R1, CPMV/HRV-R2 and CPMV/HRV-R3 were produced. For comparative studies, a chimera, CPMV/HRV-44–45$_1$, was made by insertion of the peptide between Asp44 and Asp45 in the βC'-βC" loop of the S protein (Taylor et al. 2000).

These chimeras were genetically stable as demonstrated by RT-PCR and sequencing after multiple passages and the yields of virions were generally similar to those obtained with wild type virus, with the exception of CPMV/HRV-L1 where the yield was reduced to approximately half of the wild type level. Protein sequencing and SDS PAGE of purified CPMV/HRV-II, CPMV/HRV-R1, CPMV/HRV-R2, CPMV/HRV-R3 and CPMV/HRV-44/45$_1$ showed there was a protease cleavage at the C-terminal end of the insertion, although the inserted peptides were still associated with the capsid, as demonstrated by ELISA and structural analysis (Lin et al. 1996; Porta et al. 1994). CPMV/HRV-L1 was an exception. Electrophoretic analysis indicated that repositioning the epitope one amino acid to the left of its original position in CPMV/HRV-II (i.e., between Pro21 and Ala22 rather than between Ala22 and Pro23 of the S protein) in CPMV/HRV-L1 substantially inhibited the cleavage reaction (Taylor et al. 2000).

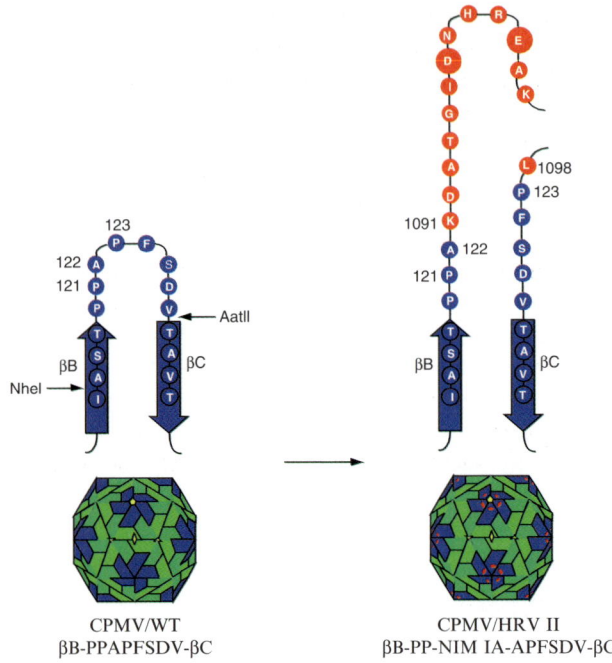

Fig. 3 Epitope presentation in CPMV chimeras. The *upper part* of the figure illustrates residues in the βB-βC loop of CPMV and the location of restriction sites as well as the residues in the chimera and the position of the spontaneous cleavage in the chimera; the *lower part* illustrates the location of the exposed loop of the chimera on the particle surface with the insertion represented in *red*. There are two unique restriction enzyme sites, Nhe I (natural) and Aat II (prepared by site-directed mutagenesis), in the region encoding the βB-βC loop of the small subunit in the infectious clone. These two sites are used in the cassette mutagenesis. Oligonucleotides are introduced to restore the native sequence of those residues and to insert the NIm-IA antigenic sequence of HRV14 between Ala122 and Pro123. Two residues, D1091 and E1095 (numbering from HRV14), which were important for the NIm-IA immunogenicity, are drawn as *larger circles*. The chimera thus generated has the foreign sequence presented at the pentamers of the virus capsid (*lower right*). The majority of the virus isolated is fragmented between K1097 and L1098. (Lin et al. 1996, with permission)

Imaging of CPMV Chimeras by X-Ray Crystallography

Since CPMV/HRV-II, -R1, -R2, and -R3 are fairly similar in terms of positioning and cleavage of the inserted peptide, crystallographic and immunological studies of CPMV chimeras with NIm-1A insertion focused on a comparison of CPMV/HRV-II, CPMV/HRV-L1 and CPMV/HRV-44–45$_1$. CPMV/HRV-II and CPMV/HRV-44–45$_1$ were crystallized isomorphously with the native CPMV in the I23 space group and the crystal structures were determined by improving the phases, calculating from the native structure through averaging (Lin et al. 1996; Taylor et al. 2000). CPMV/HRV-L1 was crystallized in the R3 space group and the structure was determined by molecular replacement with the structure of the native virus as the initial phasing model for phase refinement (Taylor et al. 2000).

The NIm-1A antigenic sequence adopted a smooth conformation on the surface of CPMV/HRV-II particle, in contrast to the convoluted conformation in the native environment, and displayed high temperature factors indicating high mobility, which was attributed to the freedom resulting from the cleavage of the peptide (Fig. 4) (Lin et al. 1996). Not surprisingly, the peptide on the surface of CPMV/HRV-44–45$_1$ showed a similar tendency in adopting extended conformations, only with multiple conformations that could only imaged at lower resolution, probably because of even higher mobility (Figs. 5 and 6) (Taylor et al. 2000).

The crystal structure of CPMV/HRV-L1 showed the continuous density for the peptide in agreement with the biochemical analysis that the insert is not cleaved. Thus the epitope is now presented as a closed loop (Figs. 5 and 6), rather than as a sequence with a free C-terminus, as found with CPMV/HRV-II and CPMV/HRV-44/45$_1$. However, despite this improvement in the mode of presentation, the

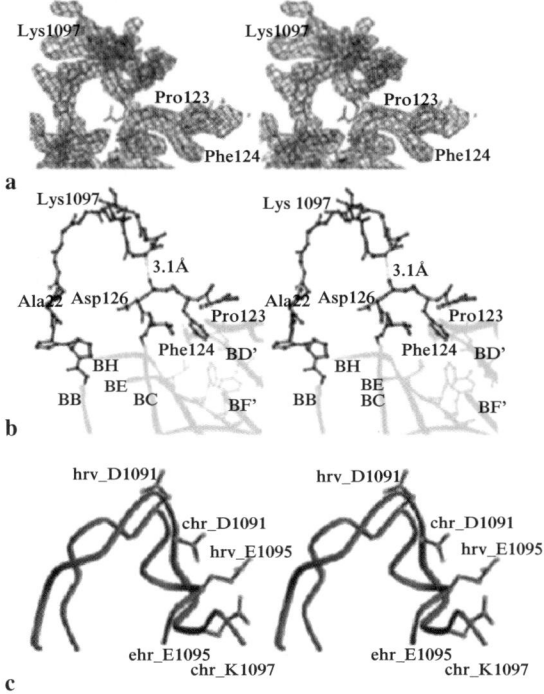

Fig. 4 a Electron density for the chimeric CPMV particles expressing the HRV14 NIm-1A site in the βB-βC loop of the S protein. **b** The model fitted the density. **c** A stereo view comparing the chimera loop with the native NIm-IA loop of HRV14 (the more convoluted structure); D1091 and E1095, which defined the NIm-IA epitope in that changes to either of these residues stopped the virus from being neutralized by specific monoclonal antibodies, are labeled in both loops. The NIm-IA loop displayed three turns. (Lin et al. 1996, with permission)

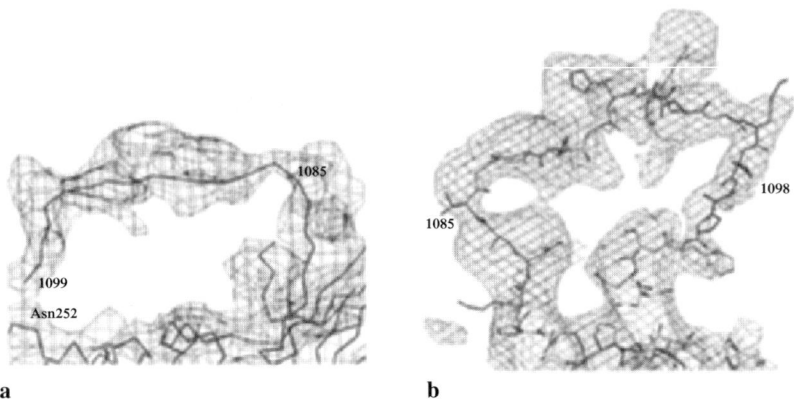

Fig. 5 Electron density maps of CPMV/HRV chimeras with the HRV14 NIm1A site inserted at different positions. **a** CPMV/HRV44/45$_1$. Electron density that was modeled for the extended conformation of insertion. The electron density is in chicken wire and the model is made with Cα tracing in *black*. The length of the density can accommodate all the inserted residues, plus a residue (1099) at the end of the insertions, in agreement with the biochemical analysis. A break of the density was obvious at the position where the cleavage occurred and the new C-terminus interacted with Asn252 of the large subunit. The electron density is contoured at 1σ. **b** CPMV/HRV-L1. Electron density that was modeled for the inserted loop. The density in chicken wire is continuous and all the residues (1085–1098) of the inserted peptide can be fitted (in *black*). The density is contoured at 2σ. (Taylor et al. 2000, with permission)

native structure of the NIm-1A sequence could not be fitted into the electron density of CPMV/HRV-L1, demonstrating the necessity of modifying the surrounding environment to achieve conformation-dependent peptide presentation (Taylor et al. 2000).

CPMV Chimeras as the Vaccine Candidates

Antisera raised in rabbits against purified virions of the three HRV chimeras that had been investigated crystallographically were used to investigate the influence of the different modes of presentation on the immunological properties of the inserted peptides. To measure the level of anti-HRV-14 antibodies in the sera, their ability to bind to native HRV-14 was tested by antigen coated plate ELISA (Fig. 7). The results showed that antibodies raised against CPMV/HRV-II and CPMV/HRV-44–45$_1$ chimeras, despite reacting strongly with HRV-14 VP1 in Western blots, bound poorly to intact HRV-14 particles. Indeed, their binding curves differed little from that obtained with wild type CPMV, suggesting that the limited binding observed might be nonspecific. However, the binding of antibodies raised against CPMV/HRV-L1 to HRV-14 particles was greatly enhanced compared with that of antibodies raised against the other two CPMV/HRV chimeras. Since both CPMV/HRV-II

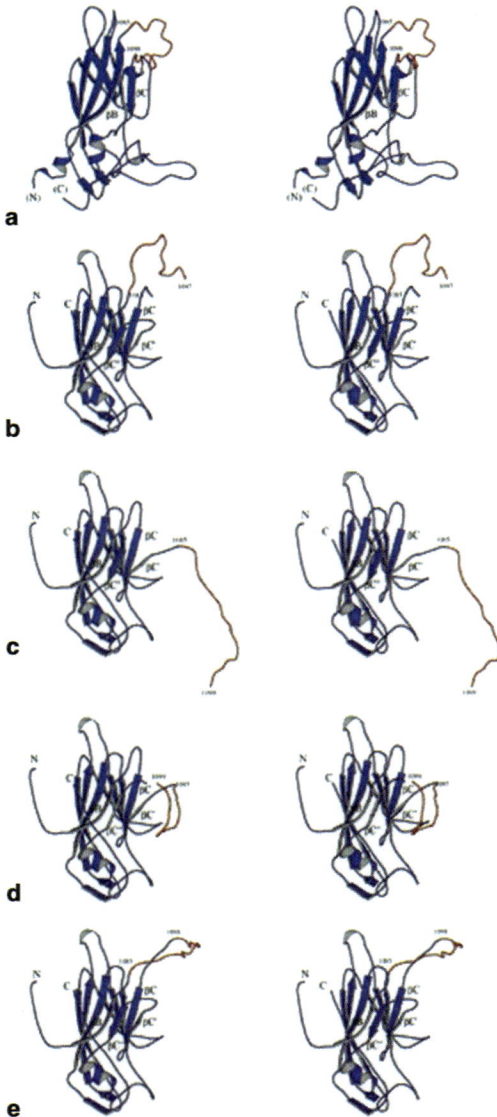

Fig. 6 Stereo views of NIm1A sequences presented on the viral surfaces. The NIm1A sequences are in *red*. **a** VP1 of HRV14. The sequence in its native environment. The N- and C-termini of VP1 are truncated in this presentation. **b** CPMV/HRV-II.A structure. The peptide is folded as a pseudo-loop bonded by a noncovalent hydrogen bond. **c** The extended conformation of the insert in CPMV/HRV44/45$_1$. The peptide extends as far as the L subunit and its C-terminus interacts with Asn252 of the L subunit. **d** Folded insertion in CPMV/HRV44/45$_1$. The insertion folds back and interacts with the N-terminus of the βC″ loop with its C-terminus, as if the cleavage did not occur. **e** CPMV/HRV-L1. The insert is extended as far as possible and its conformation is still unlike that in its native environment. (Taylor et al. 2000, with permission)

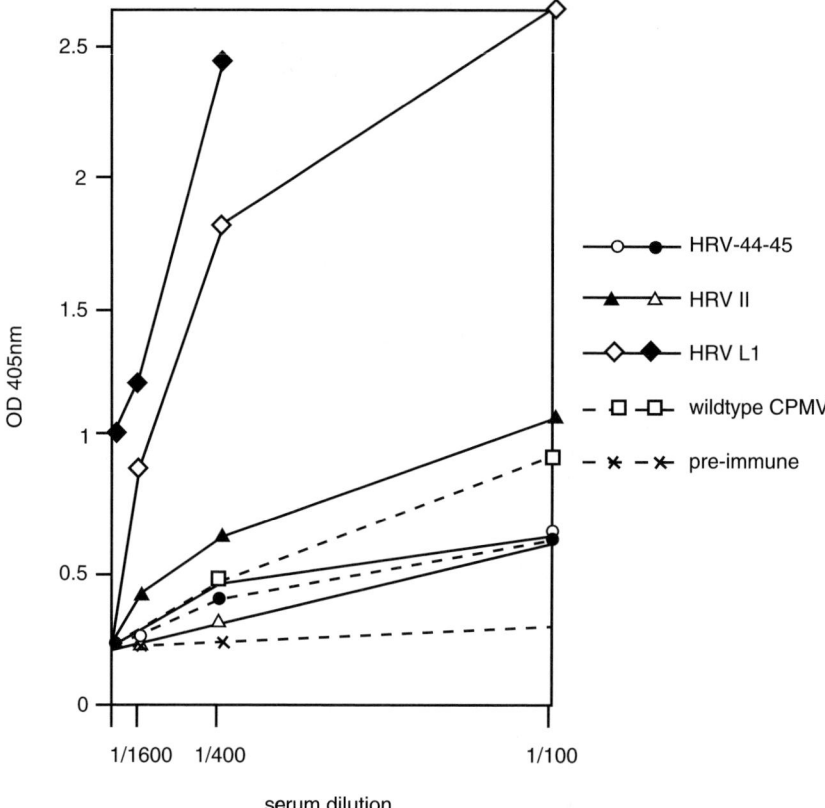

Fig. 7 Recognition of HRV-14 in ACP-ELISA by antisera produced in rabbits against CPMV wild type, HRV-44–45$_1$, HRV-II and HRV-L1. Reactivity of a preimmune serum is also shown. Binding was detected by the use of alkaline phosphatase-conjugated goat-anti-rabbit antibodies and p-nitrophenyl phosphate. The resultant OD$_{405nm}$ is shown on the y-axis. (Taylor et al. 2000, with permission)

and CPMV/HRV-44–45$_1$ presented the NIm-1A site as a peptide free at its C-terminus, while CPMV/HRV-L1 displayed it as a closed loop, these observations indicated that the structural constraint of the HRV-14 peptide played an important role in its immunological properties. In spite of their improved binding properties, the sera raised against CPMV/HRV-L1 were, like those raised against CPMV/HRV-II and CPMV/HRV-44–45$_1$, non-neutralizing. This was consistent with the observation that sequences outside that inserted into the HRV chimeras are necessary to create a fully functional NIm1A site.

This study represented the first occasion for any epitope-presentation system in which the crystal structures of foreign peptides were correlated with their immunological efficacy. The data from studies of NIm-1A sequence presented on the

CPMV surface confirmed the fact that the precise mode of presentation of an epitope on the surface of a carrier could be crucial for its immunological properties. For epitopes such as NIm-1A, which adopted a constrained structure in their native context, presentation as a closed loop is probably essential for good mimicry. The fact that the structure of the NIm-1A antigenic site on HRV-14 could still not be fitted in detail to the electron density corresponding to the insert in CPMV/HRV-L1 suggests that there might be room for further improvement of presentation. Additional constraints, such as disulfide bonds and a metal binding site, could be introduced into CPMV/HRV-L1 and the conformation of the insert could be modulated in vitro. Overall, the results demonstrated the potential of the CPMV-based epitope presentation for studying the relationship between peptide structure and immunogenicity.

Despite the problems obtaining accurate structural mimicry for peptides expressed on the surface of CPMV, there are chimeras that have been shown to be capable of eliciting protective immunity when administered to experimental animals. For example, a CPMV chimera (CPMV-PARVO1) expressing a 17 amino acid epitope from the N-terminal region of the VP2 capsid protein of *Canine parvovirus* was found to protect both mink and dogs against subsequent challenge when administered subcutaneously (Dalsgaard et al. 1997; Langeveld et al. 2001). Protection with chimeras expressing epitopes of bacterial origin have also been reported. A chimera expressing a 30 amino acid sequence from the fibronectin-binding protein of *Staphylococcus aureus* was able to protect rats against endocarditis and the serum from the rats was able to protect mice against weight loss due to *S. aureus* bacteremia (Rennermalm et al. 2001). Likewise a chimera expressing a 34 amino acid sequence from the outer membrane protein F of *Pseudomonas aeruginosa* was able to protect mice against challenge by two different immunotypes of *P. aeruginosa* in a model of chronic pulmonary infection (Brennan et al. 1999). The success of these chimeras in stimulating protective immunity probably stems from the fact the antigenic sites expressed act as linear epitopes that are active in a denatured form. Thus the accurate structural mimicry is not an essential requirement in these cases.

CPMV Particles Containing Heterologous RNA as Diagnostic Reagents

During the production of CPMV chimeras, it was noted that the modified version of RNA-2 containing the additional sequence was encapsidated into particles as efficiently as wild type RNA-2. Thus RNA-2-containing particles can accommodate RNA-2 molecules with additional lengths of heterologous sequence. Such encapsidated RNA is highly resistant to degradation. These observations have been exploited to design modified versions of RNA-2 harboring pathogen-specific sequences that can act as positive controls in highly sensitive real-time PCR-based diagnostic reactions (King et al. 2007).

Chemical Fabrications of CPMV-Based Assemblies

Though genetic assembly of antigenic devices (chimeras) is effective in generating useful vaccine candidates, there are limitations to this approach. It has not proven possible to stably express peptides larger than about 30 amino acids on the virus surface and some chimeras with even short peptides could not be recovered from plants at a reasonable level, especially if the insert had a high pI (Porta et al. 2003). The effect of pI can, however, be relieved by flanking the inserted sequence with acidic residues (N.P. Montague, C. Porta, G.P. Lomonossoff, unpublished data). It is also not practical to express full-length proteins on the virion surface and no nonproteinaceous components can be incorporated into the capsid by genetic means. Additionally, genetic insertion can only be made in specific locations and there is no flexibility in patterning the exogenous components. These limitations can be circumvented by chemical attachment strategies; small chemical moieties as well as biological peptides, and even full-length proteins can be attached to the viral scaffold using a range of bioconjugation reactions. The chemical methods complement the genetic assembly of CPMV-based devices and allow a greater variety of nanocomponents to be assembled in designed patterns.

Assembly of CPMV–Gold Conjugates by Thiol Chemistry

Thiol-selective chemical reagents have been employed to probe the reactivity of non-disulfide-linked Cys residues in the native CPMV capsid. There are no free sulfhydryl groups on the exterior CPMV surface as shown in the crystal structure (Lin et al. 1999) and the attachment of fluorescein molecules through a thiol-reactive maleimide was shown to occur only at interior sites of wild-type CPMV (Fig. 8) (Wang et al. 2002). Thiol-addressable CPMV mutants with Cys residues on the exterior surface were generated by insertion of a Cys-containing peptides (GGCGG or similar sequences), as a cassette at the sites previously used for epitope presentation (Lin et al. 1996; Taylor et al. 2000). A mutant, vEFα, with the Cys residues placed between positions 298 and 299 of the βE-βF loop in the L subunit was the most extensively characterized. Labeling of the reactive thiols can be driven to near completion, i.e., up to 60 chemical ligands can be attached and displayed on the exterior capsid surface (Wang et al. 2002).

After establishing the chemical accessibility of the introduced Cys side chains, the mutant particles were reacted with monomaleimido-Nanogold. The derivatized particles were imaged by cryoelectron microscopy and image reconstruction. Figure 9a shows the resultant structure of the CPMV-gold complex, and Fig. 9b shows a difference map in which density for the model CPMV structure was subtracted from the density computed in the image reconstruction. The gold particles were clearly visible at the positions of the inserted Cys residues (Fig. 9c), providing an example of the installation of functional structures at designated positions on the

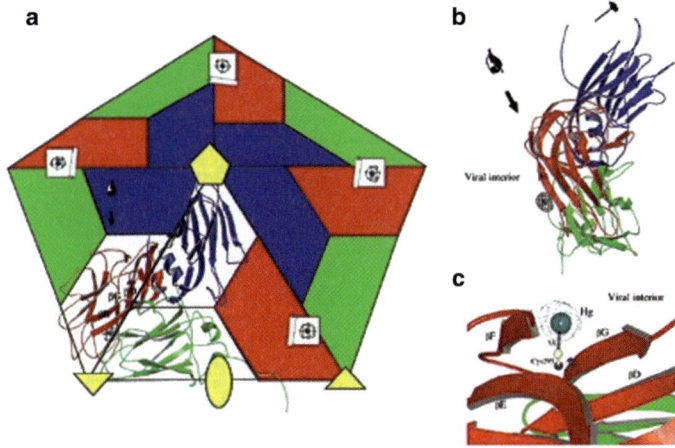

Fig. 8 Crystallographic analysis of CPMV particles derivatized with ethyl mercury phosphate. Amplitudes in the Fourier series calculation were obtained by subtracting structure amplitudes computed from the atomic model of native CPMV from the measured structure amplitudes of the EMP derivative. The difference amplitudes and native phases were used to compute electron density. **a** The pentameric assembly of CPMV protein about the fivefold symmetry axis. The difference electron density map reveals bound EMP molecules to be located solely at a single position below the outer capsid surface corresponding to Cys295; five such sites are shown here. **b** A view showing the fold of the CPMV asymmetric unit with EMP difference density. **c** A close-up view showing the position of the EMP difference density. (Wang et al. 2002, with permission)

icosahedral protein template. These experiments clearly demonstrated that a virus particle can function as a convenient and programmable platform for chemical reactions.

In order to design various nanomaterials, it is important that molecules can be attached to different sites and that materials can be assembled in different patterns; therefore, it was necessary to install various attachment sites at different locations on the viral nanobuilding block. Based on the experience gained with the Cys-added CPMV mutants produced using the chimeric virus technology, three basic design principles were implemented: first, the residues at the insertion site should not be involved in interactions with others. Second, the thiol groups should be placed in shallow pockets, which would alleviate interparticle cross-linking through disulfide linkages, while allowing the attachment of smaller molecules. Third, symmetry elements on the virus surface should be taken into account to reduce the number of altered sites.

Based on those principles, two double Cys mutants (mutants with two introduced Cys residues per asymmetric unit) were generated (Blum et al. 2005). CPMV mutant vT228C was initially made by mutation of Thr228 of the L subunit to Cys. This mutant was then used for the generation of the two double Cys mutants.

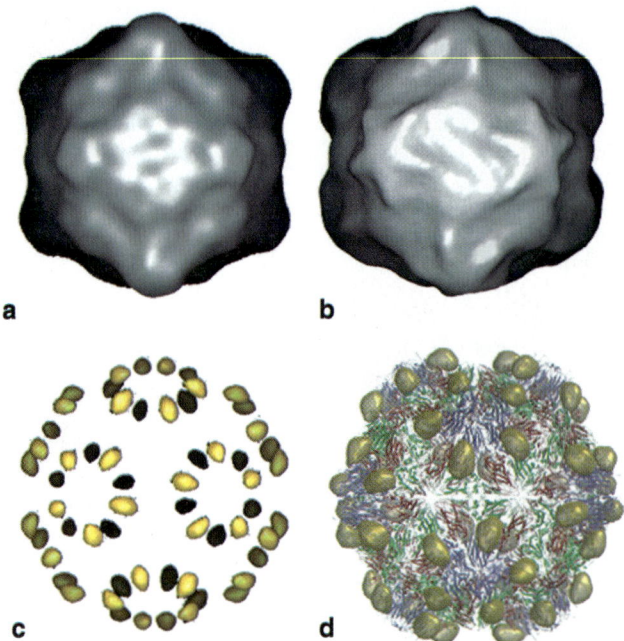

Fig. 9 Electron cryomicroscopy analysis of derivatized CPMV Cys mutant. **a** Three-dimensional reconstruction of native CPMV particles. **b** Electron density of CPMV particles derivatized with gold particles. **c** Difference electron density map was generated by subtracting density computed with the native CPMV X-ray structure from the density of the derivative. **d** The difference electron density superimposed on the atomic model of CPMV showing that the gold is attached at the site of the Cys mutation

A mutation of Asn252 to Cys generated mutant vN252C/T228C and a mutation of Thr2102 to Cys produced mutant vT2102C/T228C. As a consequence of the twofold symmetry in the virus particles, the double Cys mutations create thiol footprints with four anchoring points (Fig. 10).

The two double Cys mutants were crystallized isomorphously with the native virus in I23 space group using similar conditions to those applied previously (Lin et al. 1999). Crystals of vT2102/T228C were derivatized with platinum (II) (2,2′:6′2′-terpyridine) chloride and crystals of vN252C/T228C were derivatized with parahydroxymercuric acid. The derivatives were studied by X-ray crystallography and the structure showed binding of Pt ions at the installed Cys residues (Fig. 10).

These studies demonstrate that using structural design principles and genetic engineering are powerful techniques allowing the installation of reactive groups with exact positioning. The installed functionalities can then serve as attachment sites for the conjugation and display of heterologous materials with precision.

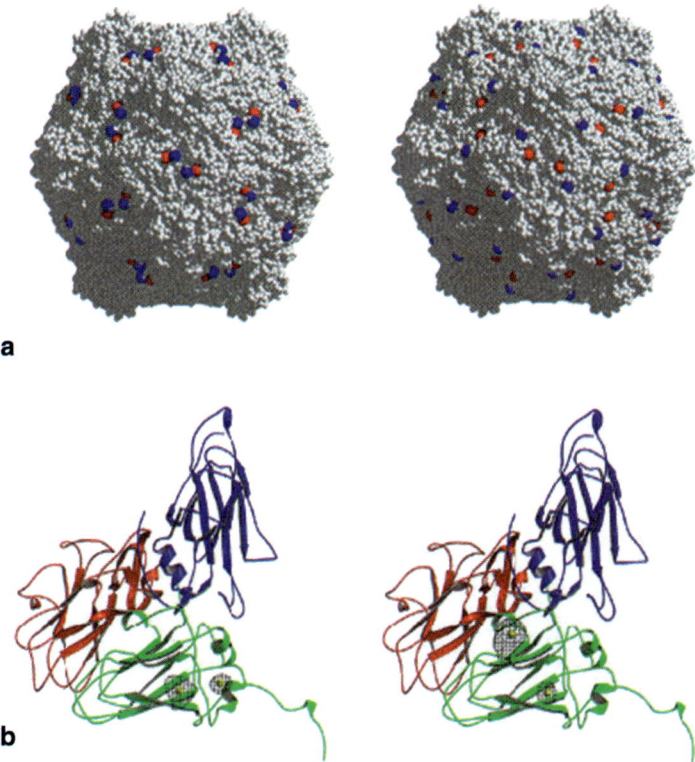

Fig. 10 a Footprints of Cys residues on the CPMV models. *Left* vN252C/T228C. Residues 252 are in *blue* and residues 228 are in *red*. *Right* vT2102C/T228C. Residues 2102 are in *blue*. **b** Ribbon diagram of CPMV asymmetric unit overlapped with different map produced from the Pt derivatives of CPMV mutants. *Left* vN252C/T228C. The virus crystals were derivatized with platinum (II) (2,2′:6′2′-terpyridine) chloride. *Right* vN252C/T228C. Virus crystals were derivatized with parahydroxymercuric acid

Engineering of Specific Amine Reactivity for Assembly of Supramolecules

For directional and controlled heterologous attachment of novel materials to the surface of CPMV, at least two types of reactive groups would be required. In addition to thiols, amine groups are also highly reactive in proteins (Hermanson 1996). Amine groups can be contributed by either the N-terminus of a polypeptide or by the side chains of Lys or Arg residues. Since the N-termini of both the coat proteins are in the capsid interior and guanidine groups of Arg are unlikely to engage in the addition reactions with *N*-hydroxysuccinimide (NHS) ester and isothiocyanate groups at pH 7 (the conditions we wished to employ), the only reactive amine groups on the virus exterior surface will be contributed by exposed Lys side chains. There are, however, five solvent-exposed Lys residues per asymmetric unit on the capsid exterior (Fig. 11), and attachment of ligands to specific single Lys side chains would be difficult, if not impossible, to achieve. Thus removal of selected Lys residues has to be carried out to eliminate the excess reactivity and to gain control over positioning. The best way of achieving this is to mutate the appropriate Lys residues to Arg residues since the polarity and positive charge of Arg side chains are likely to preserve the environment of the original Lys residues (Hermanson 1996). Table 1 lists the sequential mutants generated for the creation of Lys-minus mutants that would be associated with specific amine reactivity (Chatterji et al. 2004a).

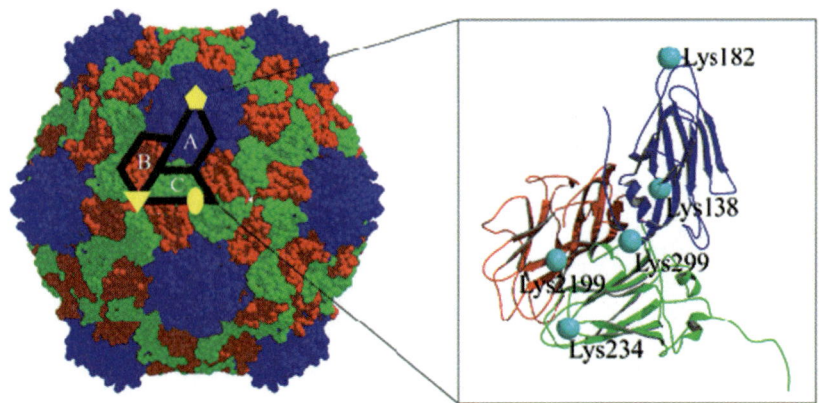

Fig. 11 *Left* Space-filling model of CPMV capsid. The reference asymmetric unit is framed and the symmetry elements are labeled. Small (S) subunits labeled A are in *blue*, and the large (L) subunits formed by two domains are in *red* (B domains) and in *green* (C domains). The *oval* represents a twofold axis; the *triangle* is a threefold axis and the *pentagon* a fivefold axis. *Right* Ribbon diagram of the asymmetric unit comprised by three jelly roll β-sandwiches with surface Lys residues represented as spheres in *cyan*. Lys138 and Lys182 are in the A domain, Lys299 and Lys234 are in the C domain, while Lys2199 is in the B domain. (Chatterji et al. 2004a, with permission)

Table 1 Lys-minus CPMV mutants

Virus	Mutation	Yield (mg/g of leaves)
Wildtype		1–1.5
vK182.K234.K299.K2199	K138R	1–1.5
vK138.K234.K299.K2199	K182R	0.4
vK138.K182.K299.K2199	K234R	1–1.5
vK138.K182.K234.K2199	K299R	1–1.5
vK138.K182.K234.K299	K2199R	1–1.5
vK182.K234.K2199	K138R/K299R	0.6
vK138.K182.K2199	K234R/K299R	1–1.5
K234.K299	K138R/K182R/K2199R	0.8
vK138	K182R/K234R/K299R/K2199R	1–1.5
vK299	K138R/K182R/K234R/K2199R	0.5
vK2199	K138R/K182R/K234R/299R	0.8

Chatterji et al. 2004a, with permission

The reactivity of the Lys-minus mutants was investigated by quantifying the absorbance at 495 nm of attached fluorescein molecules after labeling the virus with NHS ester or isothiocyanate derivatives of fluorescein (Fig. 12). Mutation of single Lys to Arg residues reduced the reactivity in each of the five cases, but none of the mutations abolished the reactivity completely (Fig. 12). In labeling reactions with fluorescein isothiocyanate (FITC), the mutation of Lys138 to Arg resulted in a reduction of about 37% of the reactivity, suggesting that Lys138 was associated with significant reactivity. Another significant reduction in the reactivity, about 26%, was observed as a result of mutation of Lys299 to Arg. Overall, Lys138 and Lys299 accounted for two-thirds of the total reactivity in the native virus, while the

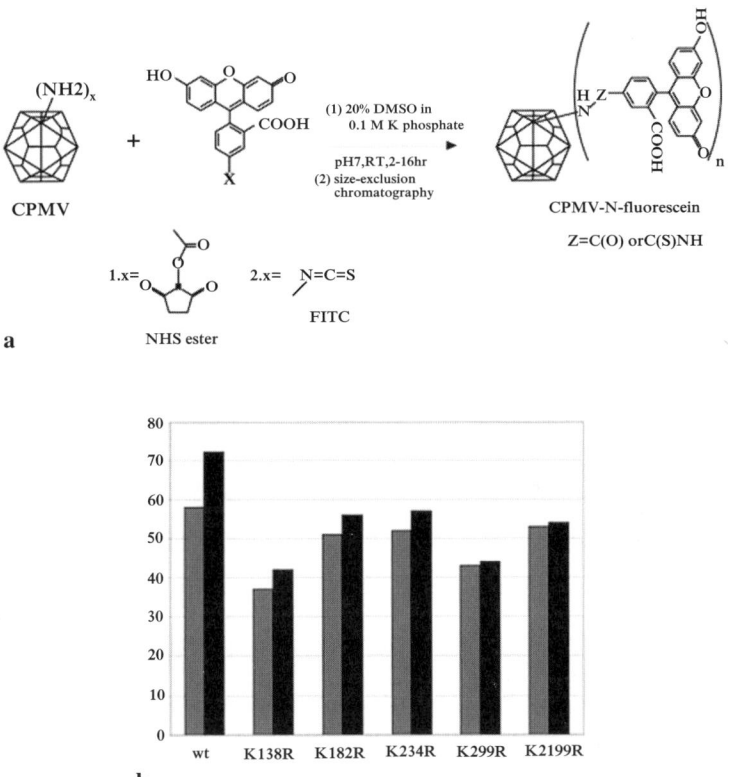

Fig. 12 Quantification of fluorescein molecules attached to CPMV. **a** Formula for the attachment of fluorescein molecules to amine groups of Lys residues by FITC and NHS ester. **b** Reduction in labeling of single Lys-minus mutants. The columns in *black* show the number of fluorescein molecules attached to the virus particles labeled with NHS ester, while the columns in *grey* denote FITC-derived virus particles. The virus mutants used in the experiment are indicated. All reactions with NHS ester were carried out at pH7.0 at room temperature for 2 h with ×200 excess dye reactants. The reactions with FITC were carried out under similar conditions but with larger excess of dye molecules (×1000) and an extension of reaction time to 18–24 h. (Chatterji et al. 2004a, with permission)

other residues, Lys182, Lys234 and Lys2199, shared roughly one-third of the remaining amine reactivity (Fig. 12 and Table 2). The derived reactivity from the five single Lys-minus mutants could account for almost all the reactivity (93.1%) of the wild type virus, indicating that the reaction was virtually specific for the Nε amines of the five surface Lys residues and that each of the exposed lysine residues reacted independently. Labeling the wild type CPMV particles and the Lys-minus mutants with fluorescein NHS ester followed similar trends to those seen with FITC, demonstrating that all surface Lys residues contributed to the reactivity (Fig. 12 and Table 2) (Chatterji et al. 2004a). Virus particles with two or three Lys mutated to Arg were also generated and derivatized with FITC. The change in reactivity due to the multiple mutations were additive when compared to that of virus particles with single Lys mutations (Table 2).

Successive mutations resulted in virus particles with different combinations of Lys residues. Two mutants, vK138 and vK299, were generated by keeping the

Table 2 Labeling of CPMV using fluorescein NHS ester and FITC

Virus (mutation)	No. of dyes/ particle	No. of dyes/ particle	Remaining reactivity[a]	Reduction in reactivity[b]
	(NHS ester)	(FITC)	(FITC)	(FITC)
Wildtype	72	58		
vK182.K234.K299.K2199 (K138R)	46	37	63.8%	36.2%
vK138.K234.K299.K2199 (K182R)	60	51	87.9%	12.1%
vK138.K182.K299.K2199 (k234R)	64	52	89.7%	10.3%
vK138.K182.K234.K2199 (K299R)	53	43	74.1%	25.9%
vK138.K182.K234.K299 (K2199R)	59	53	91.4%	8.6%
vK182.K234.K2199 (K138R/K299	32	24	41.4%	58.6%
vK138.K182.K2199 (K234R/K299R)	48	44	75.9%	24.1%
vK234.K299 (K138R/K182R/K2199R)	34	23	39.7%	60.3%
vK138 (K182R/K234R/ K299R/K2199R)	26	23	39.6%	60.4%
vK299 (K138R/K182R/ K234R/K2199R)	31	24	41.4%	58.6%
vK138[c]	183	68		
vK299[c]	185	64		

[a] $Reactivity_{mutant}/reactivity_{wild\ tye} \times 100\%$
[b] $(Reactivity_{wild\ type} - reactivity_{mutant})/reactivity_{wild\ type} \times 100\%$, which is the derived reactivity of the mutated Lys of wild type
[c] The reactions were carried out under the so-called forcing conditions that would drive the reaction to completion
Chatterji et al. 2004a, with permission

single reactive lysine residues, Lys138 and Lys299, and mutating the other four Lys residues to Arg. These two mutants were treated with with monosulfo-NHS-Nanogold to generate metal-decorated virus particles. Structural analysis of the conjugates by cryoelectron microscopy and image reconstruction showed specific labeling of the targeted Lys residues. The electron densities corresponding to the gold particles were only associated with the unique lysine residues (Fig. 13). The

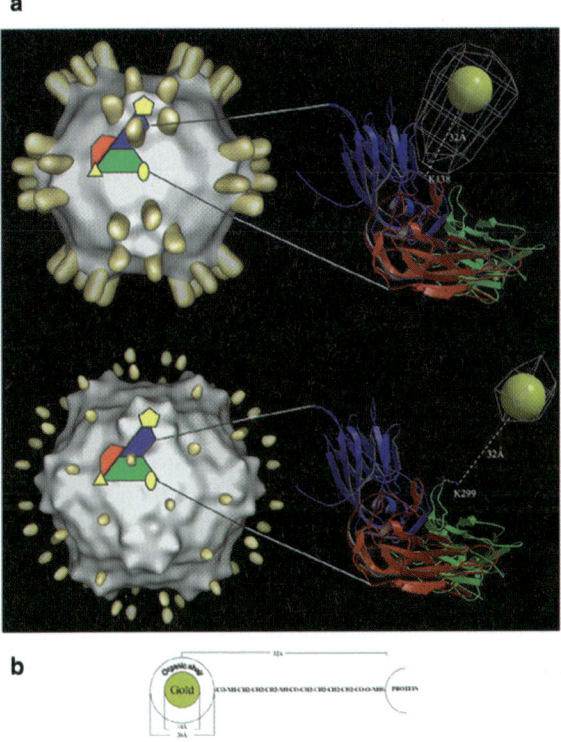

Fig. 13 a Images of gold-decorated CPMV mutants as determined by cryoelectron microscopy and image reconstruction. *Left* Cryo-EM image of gold-labeled mutant vK138. The gold particles appear as *spikes* protruding from the specific Lys residue (Lys138). *Right* Cryo-EM Image of gold/CPMV vK299 conjugate. The density corresponding to gold particles appears as *islands*, suggesting considerable latitudinal motion. Despite the higher pH of 8 employed for the reactions, which promotes the excess labeling, no labeling other than the targeted Lys residues were observed, demonstrating that labeling is specific. The images are composites of the native virus (*grey*) and difference map between the CPMV/gold conjugates and the native virus (*gold*). **b** *Top* Difference electron density, derived from vK138-gold conjugate (*left*), and the vK299-gold conjugate (*right*) superimposed with the ribbon diagram of the asymmetric unit of the virus capsid. The steric constraint labeled in vK138 conjugate restrict the movement of gold particles, while the gold particles labeled in vK299 could move more freely. The A domain is represented in *blue*, the B is shown in *red* and the C is denoted in *green*. The gold particles are also drawn as *yellow spheres* with a diameter of 14 Å. The center of the gold particles to K138 and K299 is 32 Å. The five fold axis is also shown. *Bottom* The schematic representation of the linker arm distance of 32 Å between the center of the density for gold and the labeled lysine residues, which includes the sum of the size of the gold, the organic shell around the gold and the length of the cross-linker. (Chatterji et al. 2004, with permission)

striking difference in the presentation of the gold on the two mutants suggested that the local environment of the targeted lysine residue might influence the presentation of the attached ligand. The gold particles labeled on vK138 appeared as spikes protruding radially from the virus particle, suggesting the gold particles are longitudinally mobile. In contrast, gold particles labeled on vK299 were imaged as islands of density, indicating that the gold particles are associated with latitudinal motion on a larger scale and only their vibration center is visible. K139 is located in a structural valley between the L and S subunits and only motion in a radial direction is permitted, while K299 is more on an open field, allowing more freedom of movement (Fig. 13). The decoration of CPMV with gold particles at designated locations demonstrates that the surface of the virus particle is selectively addressable; moreover, the presentation and orientation of the ligand can be tightly adjusted.

Electroactive CPMV Complexes

Display of Redox-Active Complexes on CPMV

CPMV particles were studied as a nanobuilding block for the construction of electroactive nanoscopic materials. The particles have been utilized as platforms for the display of multiple redox-active species. The motivation for these studies is derived from the fact that the particles offer a scaffold allowing the precise positioning of multiple redox-moieties. These highly symmetrical structures are expected to display interesting electrochemical properties.

Decoration of CPMV with approximately 240 ferrocenes and the attachment of 180 viologen moieties has been achieved (Steinmetz et al. 2006b, 2006c). Ferrocenes were attached using ferrocenecarboxylate and the coupling reagents N-ethyl-N'-(3-dimethylaminopropyl)carbodiimide hydrochloride (EDC) and N-hydroxysuccinimide (NHS) to accomplish facile coupling to solvent-exposed lysines on wild type CPMV (Steinmetz et al. 2006c). Electrochemical studies confirmed the presence of redox-active nanoparticles. Cyclic voltammetry showed that the CPMV–ferrocene complex displayed an electrochemically reversible ferrocene/ferrocenium couple (Fig. 14). The oxidation potential for the CPMV–ferrocene complex was shifted when compared to the free ferrocenecarboxylic acid. $E_{1/2}$ of CPMV-ferrocene was 0.23 V, and $E_{1/2}$ of free ferrocenecarboxylic acid was 0.32 V vs the Ag/AgCl electrode, respectively. This shift is expected for the conversion of the carboxyl group of ferrocenecarboxylic acid to an amide on coupling to the virus capsid, since the amide is less electron-withdrawing, and shows that the ferrocenes were stably attached to the particle surface (Steinmetz et al. 2006c).

Peak currents were measured at a range of scan rates and the linear plot of i_p μ vs $v^{1/2}$ (R=0.997) showed that the oxidation process was diffusion controlled (Fig. 14). The number of ferrocenes per CPMV particle was determined using the

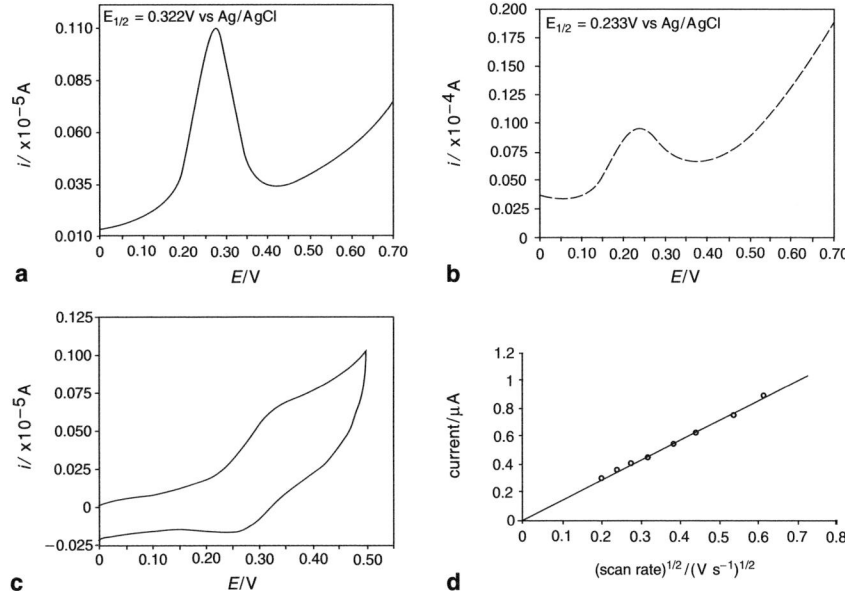

Fig. 14 a Differential pulse voltammogram of ferrocenecarboxylic acid; **b** differential pulse voltammograms of native CPMV particles (*solid line*) and derivatized CPMV-ferrocene virions (*dashed line*); **c** cyclic voltammogram of CPMV-ferrocene virions at a scan rate of 0.1 V; **d** linear plot of current vs the square root of the scan rate. (Steinmetz et al. 2006c, with permission)

Randles-Sevcik equation, and it was calculated to be around 240 ferrocene moieties per CPMV particle.

The cyclic voltammogram of the CPMV–ferrocene complexes (Fig. 14) shows a single reversible oxidation wave for the 240 redox centers displayed on CPMV. This corresponds to a simultaneous multielectron transfer at the same potential for all redox centers, and therefore the multiple moieties are independent and behave as essentially noninteracting redox units. This is an interesting and striking electrochemical property, the redox centers displayed on the insulating 30 nm in diameter viral scaffold have average distances of approximately 20–40 Å from each other (measurements are based on the crystal structure of CPMV). The fact that a simultaneous multielectron transfer of all redox units is observed can be explained by either fast rotational diffusion of the viral particle or by a kind of relay mechanism, such as electron hopping. Electron hopping between redox centers presented on the same viral particle is unlikely because the spatial separation of the redox centers is relatively large (on the electrochemical scale). However, electron hopping events between redox centers on different particles is possible, and this could contribute to the observation of simultaneous electron transfer of all 240 electrons derived from the ferrocene centers (Steinmetz et al. 2006c).

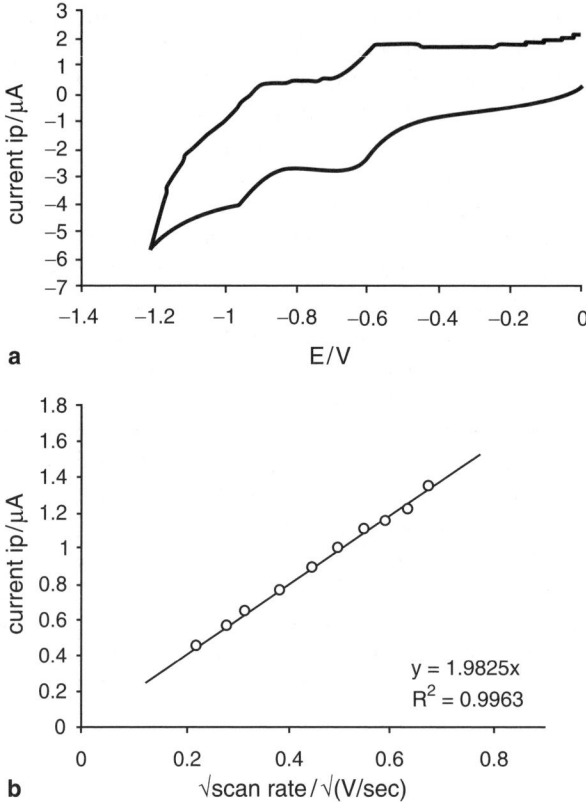

Fig. 15 a Cyclic voltammogram of CPMV-viologen complexes in aqueous buffer solution measured at a scan rate of 150 mV. b A linear plot of the measured current vs the square root of the scan rate. Electrochemical measurements were obtained and controlled using an Autolab PGSTAT 30 with GPES version 4.9 software. (Steinmetz et al. 2008, with permission)

For the attachment of viologen moieties, solvent-exposed carboxylates were chosen as attachment sites. The accessible surface profile of CPMV (the atomic coordinates for CPMV are available at http://viperdb.scripps.edu) suggests that there are eight to nine solvent-exposed carboxylate groups on the particle surface derived from aspartic and glutamic acids. To probe their reactivity the carboxylate-selective dye N-cyclohexyl-N'-(4-(dimethylamino)naphthyl)carbodiimide was reacted with wild type CPMV. UV/visible spectroscopy, native and denaturing gel electrophoresis showed that the particles could indeed be labeled with the carboxylate-selective compound (Steinmetz et al. 2006b). To facilitate bioconjugation of viologen moieties to the carboxylates on CPMV, methylaminopropylviologen [N-methyl-N'-(3-aminopropyl)-4,4'-bipyridinium diiodide] was synthesized and reacted with CPMV in the presence of the coupling reagents EDC and NHS. Cyclic voltammetric studies on viologen-decorated nanoparticles showed the characteristic two successive, one electron, reversible steps of the methyl viologen moieties

(Fig. 15) (Steinmetz et al. 2006b). The two reduction potentials, $E^{0\prime}$ −0.65 V and −0.97 V vs the Ag/AgCl electrode, for the CPMV–viologen assembly were comparable to the reduction potentials of free viologen in solution. A large difference in $E^{0\prime}$ is not expected, since the methylviologen-N-propylamine and methylviologen-N-propylamide will have similar inductive properties because of the intervening propyl group. Therefore, the attached viologen moieties behave, from an electrochemical point of view, similar to viologen in solution. The fact that a single reduction wave is observed shows simultaneous multielectron transfer of all redox units, indicating that the attached moieties behave as independent, electronically isolated units. This can be explained by the same mechanisms proposed for the CPMV–ferrocene complex: fast rotational diffusion of the viral particle, or a kind of relay mechanism such as electron hopping, or a combination of both (Steinmetz et al. 2006b). Peak currents were measured and the linear plot of i_p vs $v^{1/2}$ (R = 0.996) showed that the reduction processes were diffusion controlled (Fig. 15). The number of viologen molecules attached to each CPMV virion was estimated by use of the Randles-Sevcik equation, and it was found that around 180 viologens decorated each viral particle (Steinmetz et al. 2006b).

In summary, the feasibility of CPMV as a nanobuilding block for chemical conjugation with redox-active compounds was demonstrated. The resulting robust and monodisperse particles could serve as a multielectron reservoir that may lead to the development of nanoscale electron transfer mediators in redox catalysis, molecular recognition and amperometric biosensors and to nanoelectronic devices such as molecular batteries or capacitors.

Conducting 3D Networks on CPMV

Cys-added CPMV mutants have been used as a scaffold for a bottom-up self-assembly approach to generate conductive networks at the nanoscale (Blum et al. 2005). Two different CPMV cysteine mutants were used, one mutant displayed the cysteine residue in the βE-βF loop, the other mutant was a double mutant where the cysteines had been inserted at amino acid positions 235 and 2319, respectively. Gold nanoparticles were bound to the cysteines based on the gold–sulfur interaction (Fig. 16) (Blum et al. 2005). The attached gold nanoparticles were subsequently interconnected by molecular wires, thus creating a 3D conducting molecular network (Fig. 16) (Blum et al. 2005). To achieve this, the following molecules were used: $1,4\text{-}C_6H_4[trans\text{-}(_4\text{-}AcSC_6H_4C\equiv CPt\text{-}(PBu_3)_2C\equiv C]_2$ and oligophenylenevinylene. Molecular attachment of the linkers was confirmed by fluorescence spectroscopy. The conductance of the self-assembled molecular network was confirmed by scanning tunneling microscopy (Blum et al. 2005).

This study demonstrated once more that the CPMV building block is an excellent tool for the design and fabrication of novel materials at the nanoscale with precision. This holds great promise in assembling and interconnecting various compounds and may lead to the use of the CPMV building block for the construction of nanoelectronic systems such as nanocircuits and data storage devices.

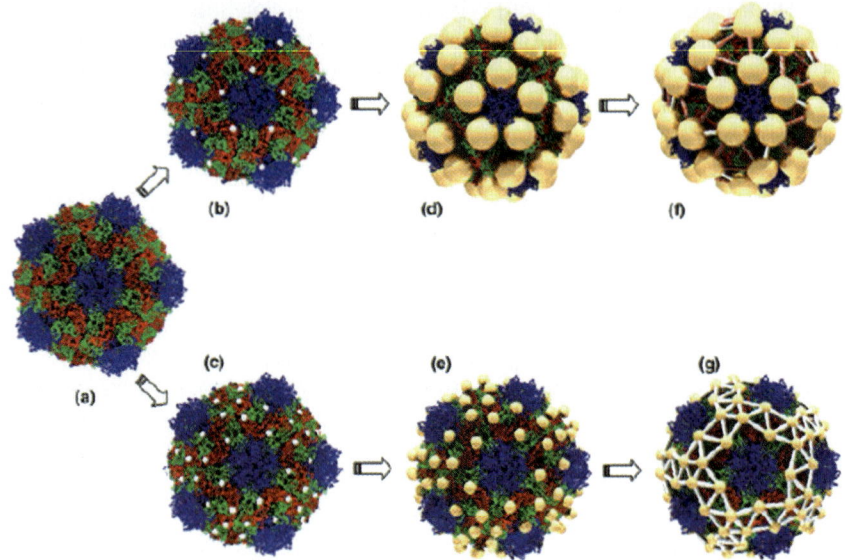

Fig. 16 Schematic of the procedure used to create molecular networks on the surface of the virus capsid. **a** CPMV capsid structure from crystallographic data. **b** EF mutant with one cysteine (*white dots*) per subunit. The four nearest-neighbor cysteine-to-cysteine distances are 5.3, 6.6, 7.5, and 7.9 nm. **c** DM mutant with two cysteines per subunit. The four nearest-neighbor cysteine-to-cysteine distances are 3.2, 4.0, 4.0, and 4.2 nm. **d** EF with 5-nm gold nanoparticles bound to the inserted cysteines. **e** DM with 2-nm gold nanoparticles bound to the inserted cysteines. **f** EF mutant with the 5-nm gold particles interconnected using di-Pt (*red*) and OPV (*silver*) molecules. **g** DM mutant with the 2-nm gold particles interconnected with OPV molecules. (Blum et al. 2005, with permission)

Targeted Delivery to Mammalian Cells by the Attachment of Functional Proteins

Chemical conjugation of proteins to CPMV complements genetic insertion methods, by allowing the attachment of peptides and proteins that would otherwise be impossible or difficult to achieve by genetic means. Different conjugation schemes were investigated to crosslink foreign proteins on the virus capsid depending upon the features of the foreign protein, and it has been shown that proteins containing Cys and Lys residues could be attached to CPMV without affecting their functions (Chatterji et al. 2004b). However, even in the absence of either of the two most nucleophilic groups on the heterologous protein, the viral scaffold can still be used for presentation of the foreign proteins. An example of such presentation is the coupling of an intron 8-encoded domain of Herstatin (Int 8) on the virus surface using thiolation reagents (Chatterji et al. 2004b).

Herstatin, an autoinhibitor of the ErbB family of receptor tyrosine kinases, is a product of an alternate splicing event from the Her2/neu gene. Her-2 intron 8 is retained and a translational readthrough results in a truncated protein consisting of subdomains I and II of the HER-2 receptor followed by a 79 amino acid C-terminal extension encoded by Int 8 (Doherty et al. 1999). Herstatin blocks receptor activation and modulates intracellular signaling through these receptors (Azios et al. 2001; Doherty et al. 1999; Justman and Clinton 2003; Molina et al. 2002). The Int 8 is a receptor-binding module that binds with nanomolar affinity to epidermal growth factor receptor (EGFR) and HER-2 receptors (Azios et al. 2001; Doherty et al. 1999). The Int 8 polypeptide was considered a candidate for conjugation to CPMV capsid as it has therapeutic potential because of its role in binding to carcinoma cells that overexpress and are driven by HER-2 or the EGFR.

The attachment of functional Int 8 to CPMV was not straightforward because it does not have any Lys or Cys residues in its sequence. In addition, the protein is soluble and stable only above pH 8 and forms oligomers at moderate concentrations. A strategy was developed to modify the protein with a thiolation reagent before conjugation with a heterobifunctional cross-linker. The active NHS ester end of (*N*-succinimidyl *S*-acetylthioacetate was reacted with the N-terminal amine in Int 8 at pH 8.0 to form a stable amide linkage (Fig. 17). The modified Int 8 now contained a protected thiol, meaning it could be stored without degradation and which precluded oligomerization of the heterologous protein. The sulfhydryl-modified Int 8 was coupled with maleimide-activated CPMV after deacetylation with an excess of hydroxylamine HCl (Fig. 17). Western blot analysis of Int 8-modified CPMV samples revealed that the L subunit was modified with the heterologous protein and the increase in size of the L subunit was consistent with the corresponding size of Int 8 attached to the virus capsid (Fig. 17).

The CPMV–Int 8 conjugate was tested in vitro for specific binding to NIH-3T3 cells expressing the HER-2 gene. As shown in Fig. 17, the CPMV–Int 8 conjugate was found to specifically associate with NIH-3T3 cells stably transfected with HER-2 but not with parental cells. Also, wild type CPMV did not show binding to either of the two cell types tested, indicating that the bound Int 8 was biologically active and responsible for specific binding to the receptor overexpressing cells. The fact that heterologous proteins displayed on the virus capsid retain their biological properties and function after conjugation suggests that cross-linking to the virus capsid may be a potential strategy to target and deliver drug molecules into the mammalian cells (Chatterji et al. 2004b).

Fabrication of CPMV Arrays

In the previous sections, we discussed the utilization of CPMV as a nanoscaffold for attachment and positioning of inorganic and biological materials in a defined and precise fashion. Besides serving as a template for arraying other materials, the particles themselves can be self-assembled into regular 2D and 3D arrays.

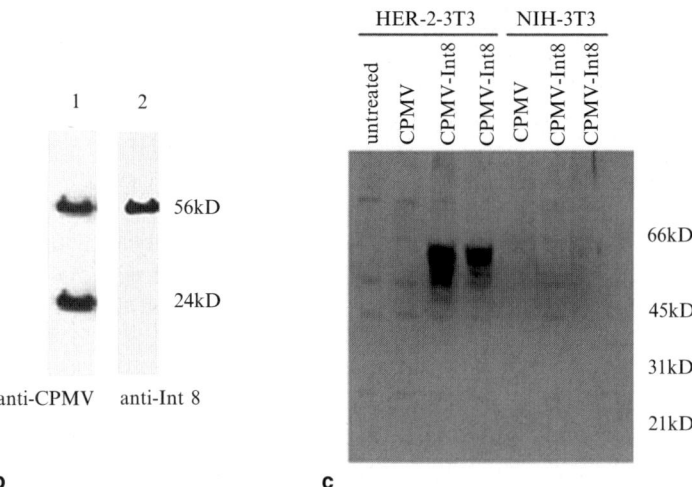

Fig. 17 a *Top* Reaction scheme for generating a thiol group on the Int 8 using the thiolation reagent SATA. (*i*) The active NHS ester end of SATA reacts with the N-terminal amine of Int 8 to form a protected sulfhydryl derivative and (*ii*) deprotection with hydroxylamine of the acetylated thiol of SATA modified Int8 yields a free sulfhydryl group that can be used for conjugation to CPMV using SMCC. *Bottom* Detection of Int 8-CPMV conjugate by Western blotting. The Int 8 encoded 79aa peptide was cross linked to the virus capsid via unique lysine residue present on the large subunit using a heterobifunctional cross-linker. Virus specific peak (*lanes 1–4*) was eluted from the Superose6 column and verified with anti CPMV (*left*) and anti Int 8 (*right*) antibodies. The retarded band at 56 kD indicates the increase in the molecular weight of the viral small subunit as a result of conjugation. The specificity of the reaction was judged by the absence of any modification of the small subunit with Int 8. **b** Specific binding of the CPMV–Int8 conjugate to NIH-3T3 cells stably transfected with HER 2 (17–3-1). A slow migrating band at approximately 56 kd was observed with anti-CPMV antibodies only in conjugated samples (*lanes 3 and 4*). The wild type CPMV (*lanes 2 and 5*) did not show any binding to HER-2 transfected or parental 3T3 cells, indicating that the intron 8-encoded domain is needed to direct binding to the HER-2 receptor. The untreated cells were used as a negative control (*lane 1*). The specificity of binding was also confirmed by assessing the ability of the conjugate to bind to parental NIH 3T3 (*lanes 5–7*). (Chatterji et al. 2004b, with permission)

3D CPMV Arrays via Crystallization Procedures

CPMV can be crystallized in at least three space groups (Fig. 18), leading to the formation of 3D arrays with ordering to near atomic precision. CPMV crystals, like other virus crystals, contain large solvent channels, which can be visualized by the sectioning of the crystal followed by staining with uranyl acetate and imaging by electron microscopy (Fig. 18). The potential exists to exploit these repeating channels within the CPMV crystal to allow the diffusion of nanomaterials into the crystal interior and binding to the virus particles in situ. Antibodies against CPMV were diffused into the crystals and in situ binding was observed (Johnson and Harrington 1985). The channels can also be utilized for confined and regular growth of metals such as palladium and platinum (Falkner et al. 2005). In order to allow the crystals

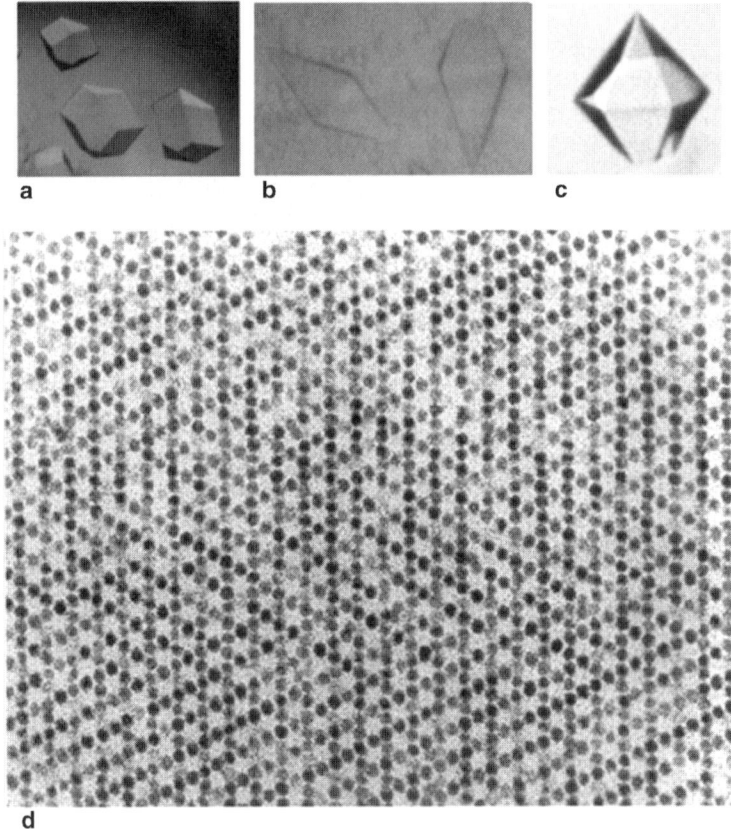

Fig. 18 CPMV crystals **a** in a cubic space group; **b** in an orthorhombic space group; **c** in a hexagonal space group. **d** EM image of thin sections of the hexagonal crystal (**C**) and stained by uranyl acetate, which shows perfect packing of the virus particles and large solvent channels in excess of 30 nm wide

to be dried or to be exposed to a variety of solvents with little adverse effects on crystal quality, CPMV particles within the crystals were interlinked using glutaraldehyde. After reinforcing the crystals, they were exposed to a buffer containing potassium tetrachloropallidate(II), followed by exposure to a buffer containing potassium tetrachloroplatinate(II) hydrate and sodium hypophosphite. The catalytic reaction led to the formation of Pd and Pt within the crystal as confirmed by imaging sections by scanning electron microscopy equipped with an energy dispersive X-ray. It was found that 10% Pd and 50% Pt by weight was formed in the samples. Transmission electron microscopy studies confirmed that the formation of the metals was confined to the channels (Falkner et al. 2005).

3D CPMV Arrays via Layer-by-Layer Assembly

Immobilizing and arraying bio(nano)particles onto solid supports is a desired requirement in biomaterials science and nanotechnology. Multilayered thin film assemblies are of growing interest for the development of miniaturized sensors, reactors, and biochips. Layer-by-layer self-assembly methods have been developed allowing the deposition of various components in a defined and ordered way. In general, CPMV particles are excellent candidates to serve as building blocks for the construction of arrays using layer-by-layer technologies. In a proof-of-concept study, CPMV particles were covalently labeled with two different ligands: biotin moieties were displayed allowing self-assembly via biospecific interaction with streptavidin, and displayed fluorescent labels served as imaging molecules (Steinmetz et al. 2006a). Attachment of the different functionalities was achieved using the design principles described in the previous sections.

Deposition and stable immobilization of CPMV particles on solid supports can be achieved using either direct chemisorption of Cys-added mutants on gold surfaces or indirect by binding biotinylated particles mediated via a thiol-modified streptavidin. Now, in order to sequentially follow the layer-by-layer assembly of the particles, different color-labeled sets of biotinylated particles were made; one set of particles was labeled with AlexaFluor 488, another set with AlexaFluor 568. The integrity of chemically modified CPMV particles was verified by electron microscopy and native gel electrophoresis. The latter in combination with UV/visible spectroscopy and dot blot tests also confirmed attachment of both functional groups, biotin and AlexaFluor moieties (Steinmetz et al. 2006a).

After the chemical composition of the building blocks was verified, arrays using layer-by-layer assembly were fabricated. First, the following CPMV bilayer was made: streptavidin–CPMV-biotin-A488–streptavidin–CPMV-biotin-A568, and vice versa. Fluorescence microscopy imaging of the CPMV arrays was consistent with successful binding of both viral nanobuilding blocks (Fig. 19). The fluorescent viral particles were spread evenly over the whole surface and dense coverage was achieved. The overlaid image demonstrated that the individual images line up well, indicating that the virions were sitting on top of each other and that a CPMV bilayer was

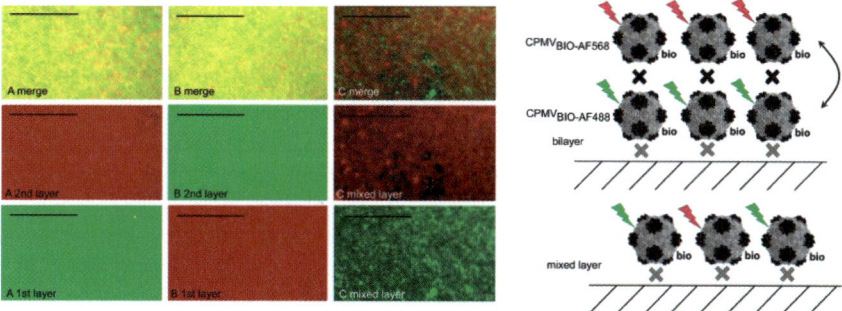

Fig. 19 Bilayers and a mixed monolayer of biotinylated (bio) and fluorescent-labeled CPMV particles on Au slides imaged via fluorescence microscopy (*left*), and diagrammatic representation of layer structures (*right*). The *green* and *red flags* show the AlexaFluor dyes AF488 and AF568, respectively. The *black cross* depicts Streptavidin (SAv); the *gray cross* shows a thiol-modified SAv. The scale bar is 10 ím. **a** bilayer of CPMVBIO-AF488 and CPMVBIO-AF568, CPMVBIO-AF488 in the first and CPMVBIO-AF568 in the second layer; merge shows the overlaid images from the first and second layer. **b** Bilayer of CPMVBIO-AF568 and CPMVBIO-AF488. **c** Mixed monolayer of CPMVBIO-AF488 and CPMVBIO-AF568. (Steinmetz et al. 2006a, with permission)

indeed formed. To further support these observations, a mixed layer consisting of streptavidin–CPMV-biotin-A488/CPMV-biotin-A568 was assembled and analyzed in the same way (Fig. 19). In this case, the particles compete for the same binding sites, which resulted in less dense and less evenly distributed coverage. The merged images do not line up, consistent with the particles occupying the same layer and competing for the same binding sites. The comparison of the overlaid image from the bilayers to that of the mixed layer further supports the successful controlled fabrication of a bilayer consisting of different fluorescent CPMV particles (Steinmetz et al. 2006a).

The construction of a trilayer via incorporation of biotinylated Cys-added mutants in the first layer was also achieved, leading to the following assembly: Cys-CPMV-biotin–streptavidin–CPMV-biotin-A488–streptavidin–CPMV-biotin-A568 (Steinmetz et al. 2006a). In theory, any number of CPMV layers can be assembled using the methods described above.

In order to make use of the CPMV arrays for materials science or nanotechnological applications, it is important to understand the assembly mechanisms and to gain insights into the mechanical properties of the assembled structures. To study this, CPMV layer assembly was followed in real time and in situ using quartz crystal microbalance with dissipation monitoring (QCMD) (Steinmetz et al. 2008). QCMD is an analytical surface-sensitive technique that is based on an acousto-mechanic transducer principle. By measuring shifts in resonance frequency (Δf) and dissipation (ΔD) of an oscillating sensor crystal, information about the deposited mass (Δm) and viscoelastic properties can be gained.

Multilayer build-up of various CPMV building blocks modified with varying densities of biotin molecules attached via longer and shorter linkers was investigated. The following CPMV building blocks were used: CPMV-(LCLC-biotin)$_F$,

CPMV-(LCLC-biotin)$_P$, CPMV-(LC-biotin)$_F$, and CPMV-(LC-biotin)$_P$; with LCLC being the long linker of about 3.50 nm, LC the short linker of about 2.24 nm, F full decoration (i.e., up to 240 biotin labels were attached), and P partial decoration, 30–40 biotins were attached. Arrays were fabricated by alternating deposition of streptavidin and biotinylated CPMV particles, and self-assembly was followed in situ and in real-time by QCMD (Fig. 20) (Steinmetz et al. 2008). The mode of multilayer assembly was different for each array fabricated. The general trend was a more regular and densely packed array arrayed when CPMV particles displaying a high number of biotin labels attached via the longer linker were assembled. In this case, a similar amount of CPMV particles was deposited at each deposition step (based on Δf). Further negative changes in ΔD upon adding streptavidin onto CPMV indicated the formation of a rigid structure (Table 3). It is expected that the multivalent display of biotin labels on CPMV in combination with the four binding

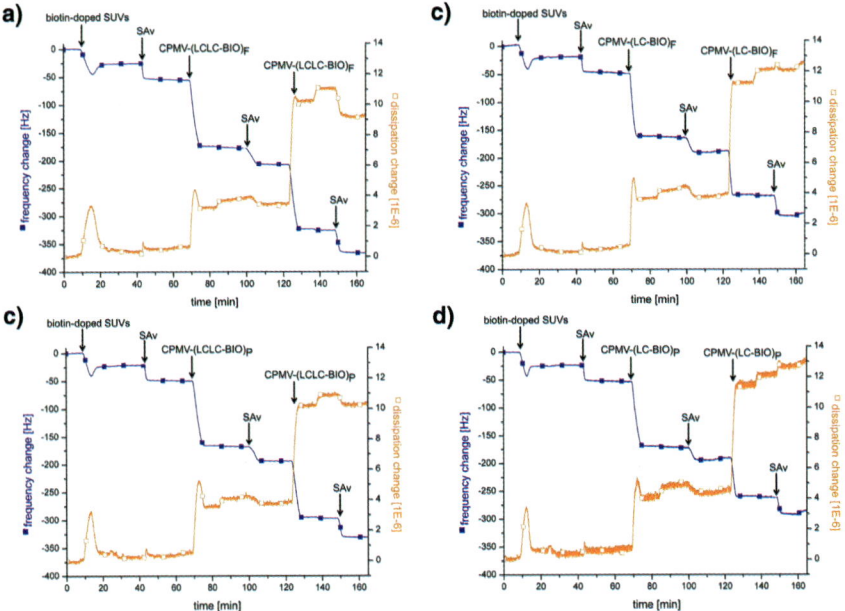

Fig. 20 Buildup of arrays consisting of alternating biotinylated *Cowpea mosaic virus* (CPMV) particles and streptavidin (SAv) on biotin-doped supported lipid bilayers (SLBs), as followed by quartz crystal microbalance with dissipation monitoring (QCMD). CPMV samples displaying two different densities of biotin molecules attached by two different linkers have been used. **a** CPMV particles displaying up to 240 biotin molecules attached via a long linker of length 30.5 Å (LCLC); CPMV-(LCLC-BIO)$_F$, with *F* full decoration. **b** CPMV particles fully decorated with biotin molecules that were attached via a short linker with a length of 22.4 Å, CPMV-(LC-BIO)$_F$. **c** CPMV particles displaying 30–40 biotin molecules attached via the LCLC linker, CPMV-(LCLC-BIO)$_P$, with *P* partial labeling. **d** CPMV-(LC-BIO)$_P$. Changes in frequency and dissipation are displayed in *blue (filled rectangles)* and *orange (open rectangles)*, respectively. (Steinmetz et al. 2008, with permission)

Table 3 QCMD responses at saturation for the alternative adsorption of streptavidin (SAv) molecules and biotinylated *Cowpea mosaic virus* (CPMV) particles

	(LCLC-BIO)$_F$[a]		(LC-BIO)$_F$[a]		(LCLC-BIO)$_P$[a]		(LC-BIO)$_P$[a]	
	Δf (Hz)	ΔD (10^{-6})	Δf (Hz)	ΔD (10^{-6})	Δf (Hz)	ΔD (10^{-6})	Δf (Hz)	ΔD (10^{-6})
SAv (1st layer)	−29	0.2	−29	0.2	−29	0.2	−29	0.2
CPMV (2nd layer)	−121	3.1	−114	3.6	−118	3.6	−118	4.1
SAv (3rd layer)	−29	−0.4	−26	−0.6	−26	−0.4	−22	−0.5
CPMV (4th layer)	−119	7.7	−82	8.2	−103	7.0	−70	7.5
SAv (5th layer)	−40	−1.8	−37	−0.0	−34	−0.7	−30	0.5

[a]*LC* short linker; *LCLC* long linker; *P* partial decoration; *F* full decoration.
Steinmetz et al. 2008, with permission

sites of streptavidin for biotin promote interlayer and intralayer cross-linking when the array is densely packed.

In stark contrast were the mechanical properties of arrays formed with CPMV particles displaying a low number of biotins attached via the shorter linker. Here, the second virus layer only accounted for 60% of the coverage that was achieved in the first virus layer (Table 3). In addition, particle cross-linking occurred to a negligible degree in the first CPMV layer and was not evident in the second layer. The formation of a viscoelastic with less dense particle packaging was indicated (Steinmetz et al. 2008).

The observation that the mechanical properties of the CPMV arrays can be tightly tuned by adjusting spacer length and density is interesting and important. The judicious use of appropriate linkers will allow the design of thin films with the desired needs and properties: rigid and densely packed vs viscoelastic and less densely packed. The ultimate goal of designing viral nanoparticle multilayer assemblies is to fabricate functional assemblies. The CPMV building blocks aligned into layers can serve as templates for the symmetrical and regular alignment of functional molecules. In the proof-of-concept experiment, we incorporated fluorescent labels into the layers that served as tags for image analyzes (Steinmetz et al. 2006a). However, any other chemical or biological molecule could replace the dyes.

2D CPMV Arrays

The first approaches toward the fabrication of a functional film were achieved by assembling quantum dots onto a 2D CPMV monolayer (Medintz et al. 2005). Two strategies were used to facilitate immobilization of CPMV and the inorganic nanocrystals: first, assembly via biotin–streptavidin interactions as described above, and second, assembly by metal affinity coordination chemistry. CPMV His-mutants expressing solvent-exposed hexahistidine sequences on the capsid exterior were made using genetic design principles (Chatterji et al. 2005). Specific immobilization of the His-mutant particles can be achieved by deposition onto NTA-functionalized templates. CPMV particles were immobilized using either strategy; quantum dots were incorporated via biotin–streptavidin interactions. Fluorescence image analyses showed that the 2D array was efficiently decorated with the luminescent quantum dots (Medintz et al. 2005). The assembly of the quantum dots on the CPMV template has advantages over a quantum dot monolayer assembled on a flat support. Not only does a CPMV-covered surface have a larger surface area compared to a flat template, but CPMV also offers defined, prearranged, symmetric, and fixed chemical attachment sites, thus allowing deposition of the crystals with high precision.

Besides fabricating continuous mono- or multilayers, single virus particle arrays have also been constructed (Cheung et al. 2003, 2006; Smith et al. 2003). Techniques such as microcontact patterning or scanning- and dip-pen nanolithography can be exploited to generate patterns of functional groups on solid supports that can then be used for binding of the viral nanoparticles. For example, lines consisting of single CPMV particles were "written" using dip-pen nanolithography. The "ink" was an alkanethiol, which was deposited on a gold "paper" by an AFM on contact mode to generate chemical templates. A Cys-CPMV mutant was then chemoselectively attached on the chemical template patterned with maleimido functionality (Smith et al. 2003).

Perspective

A new field is emerging at the exciting virus–chemistry interface. This interdisciplinary area of research is rapidly evolving and rapid progress has been made in recent years. This is not only due to recent advances in imaging and other surface-specific techniques, but also to collaborations between virologists, chemists, physicists, and materials scientists. The anticipated applications of the viral nanobuilding blocks are manifold and range from nanoelectronics to biomedical applications.

There is especially great potential for use of CPMV in biomedicine. In vitro targeting of the particles to specific mammalian cell receptors (Chatterji et al. 2004b) has already been achieved. The availability of naturally occurring empty particles or particles made by alkaline hydrolysis of the RNA (Ochoa et al. 2006) opens the door for the design and development of "smart" targeted therapies (Fig. 21).

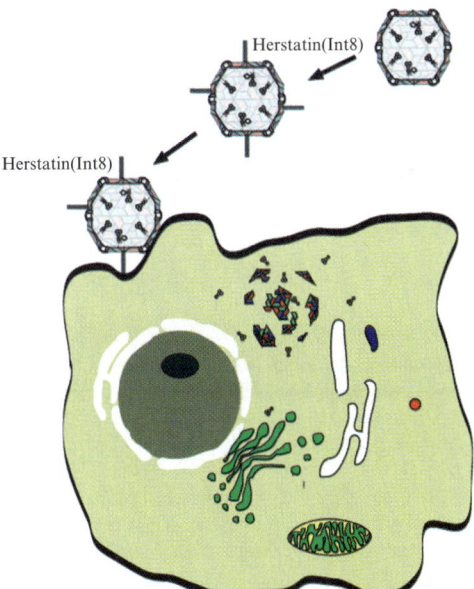

Fig. 21 CPMV for drug delivery. Therapeutics is loaded in the capsid interior and homing domains (Int 8) are attached for the targeting of cancer cells. Once the virus particles are attached to the cells, they will be endocytosed and the drugs will be released once the virus particles are degraded in the cytosol

Acknowledgements The contribution from our co-authors in previous publications is gratefully acknowledged: C. Amsinck, P.J. Barker, G.J. Belsham, A.S. Blum, E. Bock, R.S. Boshuizen, J.K. Bosworth, T.L. Brower, G. Calder, J.A. Camarero, J.I. Casal, A. Chatterji, Z. Chen, C.L. Cheung, S.W. Chung, G. Clinton, V.L. Colvin, J.B. Dai, K. Dalsgaard, J.T. Dessens, J.J. De Yoreo, K. Ebert, D. Evans, D.J. Evans, J.C. Falkner, C.E. Farrance, K.C. Findlay, M.G. Finn, P. Franzon, P. Ghosh, R.C. Gergerich, H.A. Giesing, W.D. Hamilton, M. Harrington, S. Hok, C.L. Holness, T.D. Jones, J.P. Jukes, S. Kamstrup, D.P. King, J.H. Konnert, J.P. Langeveld, K. Lee, L. Liu, J. Loveland, H. Mattoussi, A.J. Maule, I.L. Medintz, R.H. Meloen, A. Merryweather, C.A. Mirkin, A.G. Mosser, M. Mrksich, N. Montague, W. Ochoa, M. Paine, J. Perkins, S.K. Pollack, C. Porta, B.R. Ratna, S.M. Reid, R.P. Richter, P.B. Rodgers, J.B. Rohll, S.P. Salakian, K.E. Sapsford, T. Schmidt, T.L. Schull, L. Shamieh, R. Shashidhar, J.C. Smith, C.V. Stauffacher, C.M. Soto, V.E. Spall, J.P. Spatz, L. Tang, K.M. Taylor, T.J. Trentler, M.E. Turner, T. Ueno, R. Usha, A. Uttenthal, C. Vela, Q. Wang, C.D. Wilson, S.M. Wong, B.W. Woods, F. Xu. The authors would also like to acknowledge the funding from Office of Naval Research (N00014-03-1-0632 to TL and N00014-00-1-0671 to JEJ), National Institute of Health (1R01 EBB00432-02 to JEJ and NIH CA112075 to JEJ), the EU FP5 program (QLK2-CT-2002-01050 to GPL), and the Marie Curie Early Stage Training Programme (MEST-CT-2004-504273 to NFS).

References

Azios NG, Romero FJ, Denton MC, Doherty JK, Clinton GM (2001) Expression of herstatin, an autoinhibitor of HER-2/neu, inhibits transactivation of HER-3 by HER-2 and blocks EGF activation of the EGF receptor. Oncogene 20:5199–5209

Blum AS, Soto CM, Wilson CD, Brower TL, Pollack SK, Schull TL, Chatterji A, Lin T, Johnson JE, Amsinck C, Franzon P, Shashidhar R, Ratna BR (2005) An engineered virus as a scaffold for three-dimensional self-assembly on the nanoscale. Small 1:702–706

Brennan FR, Gilleland LB, Staczek J, Bendig MM, Hamilton WD, Gilleland HE Jr (1999) A chima-eric plant virus vaccine protects mice against a bacterial infection. Microbiology 145:2061–2067

Chatterji A, Ochoa W, Paine M, Ratna BR, Johnson JE, Lin T (2004a) New addresses on an addressable virus nanoblock: uniquely reactive Lys residues on cowpea mosaic virus. Chem Biol 11:855–863

Chatterji A, Ochoa W, Shamieh L, Salakian SP, Wong SM, Clinton G, Ghosh P, Lin T, Johnson JE (2004b) Chemical conjugation of heterologous proteins on the surface of cowpea mosaic virus. Bioconjug Chem 15:807–813

Chatterji A, Ochoa WF, Ueno T, Lin T, Johnson JE (2005) A virus-based nanoblock with tunable electrostatic properties. Nano Lett 5:597–602

Cheung CL, Camarero JA, Woods BW, Lin T, Johnson JE, De Yoreo JJ (2003) Fabrication of assembled virus nanostructures on templates of chemoselective linkers formed by scanning probe nanolithography. J Am Chem Soc 125:6848–6849

Cheung CL, Chung SW, Chatterji A, Lin T, Johnson JE, Hok S, Perkins J, De Yoreo JJ (2006) Physical controls on directed virus assembly at nanoscale chemical templates. J Am Chem Soc 128:10801–10807

Dalsgaard K, Uttenthal A, Jones TD, Xu F, Merryweather A, Hamilton WD, Langeveld JP, Boshuizen RS, Kamstrup S, Lomonossoff GP, Porta C, Vela C, Casal JI, Meloen RH, Rodgers PB (1997) Plant-derived vaccine protects target animals against a viral disease. Nat Biotechnol 15:248–252

Dessens JT, Lomonossoff GP (1993) Cauliflower mosaic virus 35S promoter-controlled DNA copies of cowpea mosaic virus RNAs are infectious on plants. J Gen Virol 74:889–892

Doherty JK, Bond C, Jardim A, Adelman JP, Clinton GM (1999) The HER-2/neu receptor tyrosine kinase gene encodes a secreted autoinhibitor. Proc Natl Acad Sci U S A 96: 10869–10874

Eggen R, Verver J, Wellink J, De Jong A, Goldbach R, and van Kammen A (1989) Improvements of the infectivity of in vitro transcripts from cloned cowpea mosaic virus cDNA: impact of terminal nucleotide sequences. Virology 173:447–455

Falkner JC, Turner ME, Bosworth JK, Trentler TJ, Johnson JE, Lin T, Colvin VL (2005) Virus crystals as nanocomposite scaffolds. J Am Chem Soc 127:5274–5275

Hermanson GT (1996) Bioconjugate techniques. Academic/Elsevier, London

Holness CL, Lomonossoff GP, Evans D, Maule AJ (1989) Identification of the initiation codons for translation of cowpea mosaic virus middle component RNA using site-directed mutagenesis of an infectious cDNA clone. Virology 172:311–320

Johnson J, Harrington M (1985) Antibody binding to cowpea mosaic virus in the crystalline state. In: Laver W, Air G (eds) The immune recognition of protein antigens. Cold Spring Harbor Laboratory, Woodbury, NY, pp 169–173

Johnson J, Lin T, Lomonossoff G (1997) Presentation of heterologous peptides on plant viruses: genetics, structure, and function. Annu Rev Phytopathol 35:67–86

Justman QA, Clinton GM (2003) Herstatin, an autoinhibitor of the human epidermal growth factor receptor 2 tyrosine kinase, modulates epidermal growth factor signaling pathways resulting in growth arrest. J Biol Chem 277:20618

King DP, Montague N, Ebert K, Reid SM, Jukes JP, Schädlich L, Belsham GJ, Lomonossoff GP (2007) Development of a novel recombinant encapsidated RNA particle: evaluation as an internal control for diagnostic RT-PCR. J Virol Meth 146:218–225

Langeveld JP, Brennan FR, Martinez-Torrecuadrada JL, Jones TD, Boshuizen RS, Vela C, Casal JI, Kamstrup S, Dalsgaard K, Meloen RH, Bendig MM, Hamilton WD (2001) Inactivated recombinant plant virus protects dogs from a lethal challenge with canine parvovirus. Vaccine 19:3661–3670

Liu L, Lomonossoff GP (2002) Agroinfection as a rapid method for propagating Cowpea mosaic virus-based constructs. J Virol Meth 105:343–348

Lin T, Porta C, Lomonossoff G, Johnson JE (1996) Structure-based design of peptide presentation on a viral surface: the crystal structure of a plant/animal virus chimera at 2.8 A resolution. Fold Des 1:179–187

Lin T, Chen Z, Usha R, Stauffacher CV, Dai JB, Schmidt T, Johnson JE (1999) The refined crystal structure of cowpea mosaic virus at 2.8 A resolution. Virology 265:20–34

Lomonossoff GP, Johnson JE (1991) The synthesis and structure of comovirus capsids. Prog Biophys Mol Biol 55:107–137

Medintz IL, Sapsford KE, Konnert JH, Chatterji A, Lin T, Johnson JE, Mattoussi H (2005) Decoration of discretely immobilized cowpea mosaic virus with luminescent quantum dots. Langmuir 21:5501–5510

Molina MA, Saez R, Ramsey EE, Garcia-Barchino MJ, Rojo F, Evans AJ, Albanell J, Keenan EJ, Lluch A, Garcia-Conde J, Baselga J, Clinton GM (2002) NH(2)-terminal truncated HER-2 protein but not full-length receptor is associated with nodal metastasis in human breast cancer. Clin Cancer Res 8:347–353

Ochoa WF, Chatterji A, Lin T, Johnson JE (2006) Generation and structural analysis of reactive empty particles derived from an icosahedral virus. Chem Biol 13:771–778

Porta C, Spall VE, Loveland J, Johnson JE, Barker PJ, Lomonossoff GP (1994) Development of cowpea mosaic virus as a high-yielding system for the presentation of foreign peptides. Virology 202:949–955

Porta C, Spall VE, Findlay KC, Gergerich RC, Farrance CE, Lomonossoff GP (2003) Cowpea mosaic virus-based chimaeras. Effects of inserted peptides on the phenotype, host range, and transmissibility of the modified viruses. Virology 310:50–63

Rennermalm A, Li YH, Bohaufs L, Jarstrand C, Brauner A, Brennan FR, Flock JI (2001) Antibodies against a truncated *Staphylococcus aureus* fibronectin-binding protein protect against dissemination of infection in the rat. Vaccine 19:3376–3383

Rohll JB, Holness CL, Lomonossoff GP, Maule AJ (1993) 3'-terminal nucleotide sequences important for the accumulation of cowpea mosaic virus M-RNA. Virology 193:672–679

Smith JC, Lee K, Wang Q, Finn MG, Johnson JE, Mrksich M, Mirkin CA (2003) Nanopatterning the chemospecific immobilization of cowpea mosaic virus capsid. Nano Lett 3:883–886

Steinmetz NF, Calder G, Lomonossoff GP, Evans DJ (2006a) Plant viral capsids as nanobuilding blocks: construction of arrays on solid supports. Langmuir 22:10032–10037

Steinmetz NF, Lomonossoff GP, Evans DJ (2006b) Cowpea mosaic virus for material fabrication: addressable carboxylate groups on a programmable nanoscaffold. Langmuir 22:3488–3490

Steinmetz NF, Lomonossoff GP, Evans DJ (2006c) Decoration of cowpea mosaic virus with multiple, redox-active, organometallic complexes. Small 2:530–533

Steinmetz NF, Bock E, Richter RP, Spatz JP, Lomonossoff GP, Evans DJ (2008) Assembly of multilayer arrays of viral nanoparticles via biospecific recognition: a quartz crystal microbalance with dissipation monitoring study. Biomacromolecules 9:456–462

Taylor KM, Lin T, Porta C, Mosser AG, Giesing HA, Lomonossoff GP, Johnson JE (2000) Influence of three-dimensional structure on the immunogenicity of a peptide expressed on the surface of a plant virus. J Mol Recognit 13:71–82

Vos P, Jaegle M, Wellink J, Verver J, Eggen R, van Kammen A, R, G (1998) Infectious RNA transcripts derived from full-length cDNA copies of the genomic RNAs of cowpea mosaic virus. Virology 166:33

Wang Q, Lin T, Tang L, Johnson JE, Finn MG (2002) Icosahedral virus particles as addressable nanoscale building blocks. Angew Chem Int Ed 41:459–462

Wellink J (1998) Comovirus isolation and RNA extraction. Meth Mol Biol 81:205–209

Hybrid Assembly of CPMV Viruses and Surface Characteristics of Different Mutants

N.G. Portney, G. Destito, M. Manchester, M. Ozkan (✉)

Contents

Introduction to Hybrid Networks Using Viral Nanoparticles 60
    Nanogold-Activated Release 61
Zeta Potential Study of Viral Nanoparticles 63

Abstract There is a trend toward viral-based hybrid systems to furnish viral nanoparticles with enhanced features, for function beyond a delivery vehicle. Such hybrids have included Nanogold for microwave release, quantum dots and fluorescent moieties, to provide simultaneous imaging capabilities, and iron oxide particles for image enhancement in MRI. Other systems are the subject of ongoing and vigorous research. Nanogold surface decoration of cow pea mosaic virus (CPMV) to form NG-CPMV hybrids were explored to release fluorescent carriers using microwave energy as a model system in this presentation. Thus, emergent viral-based systems will have increasingly sophisticated architectures to provide versatile functions. Zeta potential (ZP) is a powerful tool to probe the electrostatic surface potential of biological materials and remains an untapped method for studying the interaction of nanoparticles with cells. An enormous effort is being made to study nanoparticle–cell interaction, but current throughput solutions (e.g., flow cytometry) cannot differentiate between surface-attached or endocytosed particles, while standard fluorescence microscopy is tedious and costly. CPMV-WT and other mutants (CPMV-T184C, CPMV-L189C) were studied using ZP methods and rationalized based on variations in their surface exposed residue character. Understanding such subtle changes can discretely alter the cell surface interactions due to charge affinity. Applying sensitive ZP measurements on viral nanoparticles is useful to elucidating the characteristics of the surface charge and the potential interaction modes with cell surfaces they may encounter. Thus, ZP can be a unique and efficient tool for studying cell–virus interactions and aid in development of future therapeutic strategies.

M. Ozkan
Department of Electrical Engineering, University of California, Riverside, A241 Bourns Hall, Riverside, CA 92521, USA
e-mail: mihri@ee.ucr.edu

Introduction to Hybrid Networks Using Viral Nanoparticles

Viruses have been envisioned as robust nanoparticle vectors for drug delivery, vaccines, gene and cancer therapy, by harnessing their fusogenic properties for cell receptor binding, and employing recombinant methods to create unique expression systems (Lin et al. 1996). A variety of nanomaterial systems have been proposed using viruses for their regular geometries, well-characterized surface properties, and nanoscale dimensions.

Nanoscale hybrid systems using wild type and mutant icosahedral plant cowpea mosaic virus (CPMV) and insect flock-house virus (FHV) are currently under investigation. CPMV and FHV are some of the smallest viruses extensively investigated in diverse nanostructures and are more favorable for endothelial traversal by enhanced permeability and retention (EPR) effects (Maeda et al. 2001) for tumor localization compared to commonly used adenoviruses three times their size. They are excellent candidates for nanoassembly because they have been well characterized (Andrew et al. 1992, 1993), are pH and thermally stable over a wide range of conditions, and easily extracted and purified. Numerous assemblies using viruses have included CPMV-Au for addressable nanoscale building blocks (Wang et al. 2002c), CPMV-cysteine mutants for gold attachment (Wang et al. 2002b), fluorescein bound and biotinylated CPMV using succinimide esters (Wang et al. 2002a), CPMV chimeras[1] for tailored cell targeting, virus-polymer hybrids through pegylation of CPMV for reduced immunogenicity applications (Krishnaswami 2003), TMV templates for CdS deposition (Shenton et al. 1999), and virus templates for nanophase crystal growth (Douglas and Young 1999).

Previously, we have demonstrated large-scale covalent network assembly of quantum dots (QDs) and carbon nanotubes (CNTs) with CPMV and mutant FHV hybrid systems (Portney et al. 2005). Enhancement of protein uptake was demonstrated on protein-CNT conjugates without affecting cell viability or toxicity (Wong et al. 2004), suggesting that a hybrid virus-CNT network would be endocytosed more effectively. Decoration of such viral networks with QDs would allow simultaneous imaging and delivery to monitor tumor size over time. Refinement of conjugation procedures (Khorana 1993) to network any virus class could also allow modulation of release kinetics of drugs intervirally loaded within pores, without disrupting their binding potential, while providing greater release per receptor-binding event. Networking different virus classes for multiple cancer cell targets would avoid multisite directed mutagenesis of the capsid gene for each receptor, providing a wider platform for drug delivery and targeting. An example of viral-based networking was demonstrated by M.G. Finn (Strable et al. 2004), using thermally activated and reversible aggregation of oligonucleotide-conjugated CPMV particles, which could serve to actively release a potent dosage of therapeutics upon cell docking. Conjugation with QDs or gold nanoparticles for simultaneous visualization, and decoration of multiple ligands and virus classes for enhanced specificity, are examples of how hybrid nanoparticle formulations can be synergistically used to improve the potency of viral-based therapies.

Nanogold-Activated Release

Microwave-induced active release of water-solubilized quantum dots from CPMV-Nanogold networks is presented in Fig. 1, as a relevant drug release model for therapeutics in vivo. In order to demonstrate the concept of microwave radiation active release, Nanogold was conjugated to wild type CPMV (CPMV-WT) virus, loaded with water solubilized quantum dots, and visualized by fluorescence microscopy. Using conjugation chemistries, we previously applied for synthesis of QD-virus and QD-CPMV heterostructures (Portney et al. 2005), carbodiimide-mediated condensation (Carraway and Triplett 1970) between carboxyl-expressed Nanogold, and CPMV-WT lysine primary amines were exploited for efficient coupling into composite networks. Gold particles are already implemented in hybrid delivery systems such as metal cored dendrimers (Karical et al. 2003), used as oligo-gold conjugates (Parak et al. 2003), and have been proposed as contrast agents (Chen et al. 2005). The tunable optical resonance properties of gold nanoshells (Oldenburg et al. 1999) by NIR radiation has been applied to nonspecific tumor thermal ablation in mice (O'Neal et al. 2004). Also, Nanogold hybrid systems have been taken up by bacteria using microwave-assisted electroporation (Rojas-Chapana et al. 2004). With biocompatible properties (Tshikhudo

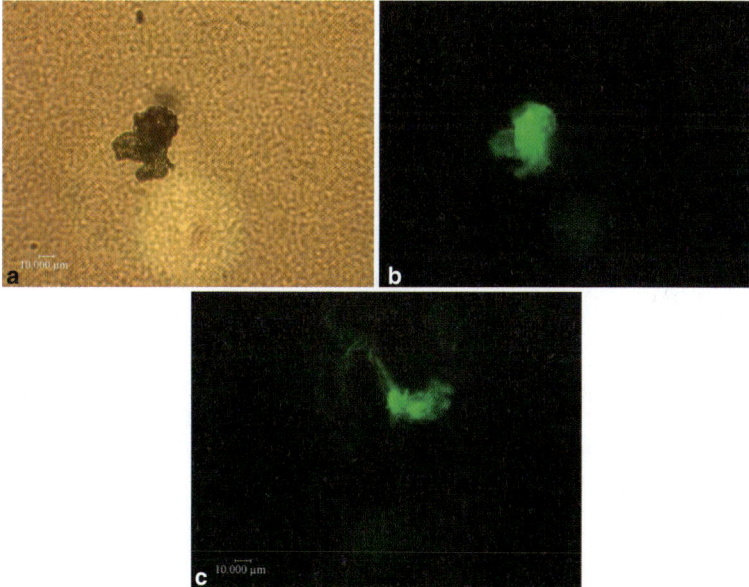

Fig. 1 **a** Bright field image mode at ×50 of QD-loaded Nanogold-CPMV networks. **b** Fluorescence mode image at ×50 of image **a**. **c** Microwave-treated NG-CPMV viral networks with release of green QD dye particles. Figure with 10-μm scale bar showing how QD particles are actively released from nanogold-CPMV networks with 10-s microwave treatment time. Characterization of QD dye fluorescence with 540-nm peak emission was obtained from an upright fluorescent microscope at ×50, using 514-nm filter and imaged using Spot diagnostic CCD. Samples were activated by exposure in a microwave oven with a 1,300-W magnetron at 2.45 Ghz

et al. 2004), optimized synthetic routes to manufacture them homogenously with a controlled number of surface functional groups (Worden et al. 2004), gold nanoparticles are useful candidates for incorporation within viral vectors for delivery applications.

Nanogold–CPMV (NG-CPMV) networks were synthesized through EDC ((1-ethyl-3-(3-dimethylaminopropyl)carbodiimide hydrochloride)) and Sulfo-NHS (N-hydroxysulfosuccinimide) -mediated condensation between carboxylated NG and abundant surface exposed CPMV-WT primary amines from lysine residues on the capsid. Negatively charged 1.4-nm-diameter gold colloidal particles (Negatively Charged Nanogold, Nanoprobes, Catalog #2023) expressed with approximately 18 carboxyl groups per particle were reacted with 300 surface-exposed amines per CPMV virion. A total of 10 nmol of stock NG solution was added to a volume of PBS buffer pH 7.38 (10 mM buffer strength), following activation by EDC (5×10^{-5} mol) for 15 min to generate a reactive O-acylisourea intermediate. A more water-soluble conjugate formed through addition of sulfo-NHS (2×10^{-9} mol) solution, and was incubated for another 15 min to extend the half-life of carboxylated NG before addition of CPMV-WT. A 306.7-mM EDC concentration with 25,000 molar excess to sulfo-NHS was used to activate the carboxylated NG solution. Addition of 1.012 mg of CPMV (5.6 MDa) and incubation with stirring for 3 h produced the final NG-CPMV networks in a final reaction volume of 500 μl. Aliquots of 100 μl were used for quantum dot loading and microwave release studies by incubation of 30 μl of stock water-solubilized quantum dots (Core-Shell CdSe/ZnS Visible Evidots, Catskill Green, Evident Technologies, catalog# 540–025), representing 0.002-mM dye solution concentration for 3 h. Samples were treated by 10-s microwave treatment (1300 W) and characterized by fluorescence microscopy. Characterization of QD fluorescence with 540-nm peak emission was obtained from an upright fluorescent microscope at ×50, using a 514-nm filter and imaged using Spot diagnostic CCD.

Microwave energy provided uniform heating of networked viral vectors and increased diffusional release kinetics of intervirally loaded drug molecules. Simultaneously, networked NG viral vectors can be monitored by MRI, combined with NIR detection of QD to confirm the biodistribution and accumulation within in vivo models, and provide contrast staining after tissue resectioning. Activated release of green QDs in Fig. 1C after 10-s treatment qualifies the extent of fluorescence distribution compared to the uniform signal in pretreated images. Usage of QDs as a therapeutic model drug provides flexible visualization capabilities with long fluorescent lifetimes, broadband absorption, and narrow emission properties (Michalet et al. 2005). Their superior photostability over conventional dyes allow simulated release studies over hours by fluorescence microscopy, whose tunable emission properties by size selection can provide NIR and visible imaging for in vivo and in vitro characterization, respectively (Kim et al. 2004). In addition, they represent the size and shape of polymer-conjugated systems in clinical use (Duncan 2003) and may simulate their diffusional release kinetics. Facile surface modification procedures enable water solubilization of QDs and biocompatible polymer encapsulation for ideal studies in vivo (Michalet et al. 2005). With the ability to limit the number of functional groups expressed on Nanogold particle surfaces, control over final network size can eventually be achieved by altering molar excess over virus and by exploiting the affinity of cysteine sulfhydryl to the gold surface through maleimide chemistry (Hainfeld et al. 1992) using cysteine-inserted CPMV mutants (Wang et al. 2002b). Since fluorescence is quenched by Nanogold, confirmation of spatial mapping of QD relative to NG-CPMV networks can be monitored during release. This model can readily be applied to monitor nonspecific targeting through EPR at tumor sites, enable time-resolved active targeting studies using the remarkable photostability of QDs, and simulate a range of drug sizes and corresponding molecular weights with comparable release behavior. Through common immunolabeling procedures (Hermanson 1996), the viral networked surface can also be modified to enable active targeting modalities.

Zeta Potential Study of Viral Nanoparticles

In order to minimize surface charge effects leading to endothelial adhesion in vivo, zeta potentials of mutants of cowpea mosaic virus were measured and compared with wild type CPMV to understand the extent of electrostatic interaction by substitution of surface exposed protein residues on the capsid. Figure 2 introduces two cysteine mutants, T184C and L189C, with residue substitutions at the small subunit coat protein, which are near the C terminus and exposed to solvent.

Zeta potential allows quantification of electrostatic potential between the surfaces of colloidal or larger size particles in solution with electrolytes, where measured average drift velocity V_s from experiment can be used to obtain the electrophoretic mobility μ_e in an electric field E according to Eq. 1.

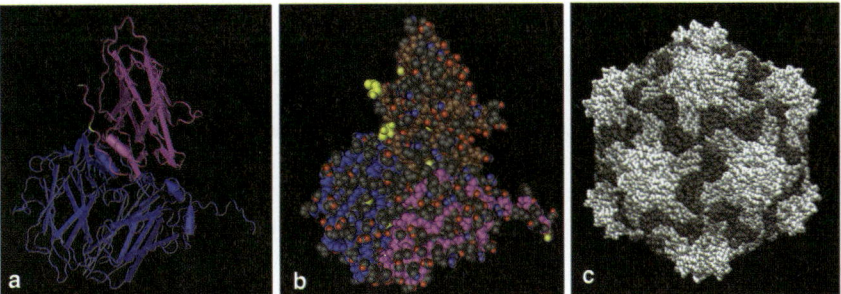

Fig. 2 a–c Molecular structures of CPMV coat proteins. **a** Ribbon diagram of CPMV capsid coat protein formed by two β-barrel domains comprised of a small (S) (*pink*) and large (L) (*blue*) subunits. Each subunit forms pentamer structures around a fivefold axis 60 times to produce the entire CPMV capsid. **b**. A space-filling model highlighting two yellow residue positions T184 for threonine and L189 for leucine at the **c** terminus, signify cysteine mutant substitutions for T184C and L189C. **c** Surface of CPMV with arrangement of capsid proteins arranged in icosahedral symmetry

$$V_s = \mu_e E$$

The distribution of ions (counterions and co-ions) in solution arrange themselves in a gradient from the charged particle surface to form a highly concentrated layer of counterions near the surface at the Stern plane, and an outer diffusive layer with greater local concentration of co-ions, forming a double layer model (Stern 1924). Zeta potential (ζ) is a measure of electrostatic potential at the surface of shear represented as the boundary between the Stern and diffuse layer. Thus, the presence of ions in solutions and their valency will strongly affect the balance of Van der Waals attraction and electrostatic and steric repulsion according to DLVO theory (Hunter 1981). A characteristic thickness representing this double layer or Debye-Hückel parameter κ^{-1}, depends on the temperature, the dielectric constant of the solvent used, and the ionic strength from free ions in solution. There are numerous double-layer theoretical models that rely on determination of mobilities to obtain zeta potentials based to on two classic limits of Smoluchowski (1903) ($\kappa a \gg 1$, $f(\kappa a,\zeta)=1.5$) and Hückel (Hückel 1924) ($\kappa a \ll 1$, $f(\kappa a,\zeta)=1$) equations, where a characteristic dimensionless electrokinetic radius κa with particle radius a is used to determine the applicability of each model in the limit of κa in Eq. 2.

$$\mu_e = \left[(\varepsilon\zeta)/(1.5^*\eta)\right] f(\kappa a,\zeta)$$

Because colloids are too large to satisfy the Hückel limit, it is best to arrange $\kappa a \gg 1$ to satisfy the Smoluchowski condition by varying solution conditions to produce a small κ^{-1}. Increasing the ion concentration decreases the double-layer thickness by an inverse proportional relationship with the square root of the ionic strength (Eq. 3). By decreasing the double layer, the distance over which repulsive forces are significant is diminished. Increasing the ion concentration too much will cause attractive forces to dominate (Fig. 3) and cause aggregation effects, leading to colloidal instability. Multivalent ions can cause such aggregation effects at much smaller concentrations. Therefore, an optimal ion concentration exists for a system of given sized particles for zeta potential measurements.

Each stock solution of CPMV viruses ranging in concentration from 3 to 5 mg/ml were originally in PBS phosphate buffer at 0.1 M at pH averaging 7.20 using monobasic and dibasic potassium phosphate. Aliquots of stock solution were added with nanopure water to make a total volume 1.50 ml to accommodate a cuvette for ZP measurement, achieving a final virus concentration of 0.04 mg/ml. Trace amounts of sodium ions from 0.002% sodium azide were also present. Each solution contained a final concentration of 1.3 mM phosphate buffer for CPMV-WT and 0.8 mM for CPMV mutants at pH 7.2. Using ZetaPALS for Zeta Potential measurements (Brookhaven Instruments Corporation, New York, NY, USA), electrophoretic mobilities were obtained through measurement of average drift velocities of virions traveling between a silver electrode through light-scattering optoelectronics and correlated to zeta

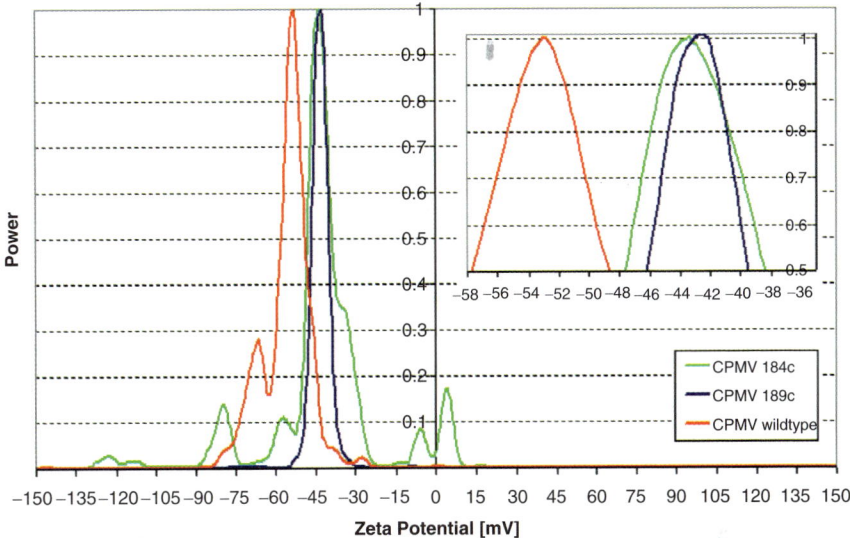

Fig. 3 Summary of CPMV zeta potential plots for wild type, mutants 184c and 189c. Best representative runs from each virus taken at peaks closest to the averages are plotted for wild type (*red*) with peak at −52.73 mV, 184c mutant in green with next smallest potential at −43.52 mV, and least negative peak for mutant 189c at −42.14 mV in purple. An inset at upper right more clearly represents a shift in zeta potential by T184C and L189C mutants compared to CPMV-WT. Each virus solution zeta potential were compiled as ten runs with five cycles averaged at a temperature stabilized 25°C with a pH of 7.20 and concentrations of 0.04 mg/ml using Ag electrodes. Zeta potential plots were calculated from measured mobility data obtained by electrophoretic migration velocities. For wild type CPMV, electrophoretic mobilities for most of ten runs were from approximately −4.7 to −3.4 mobility units with an applied electric field of 16.40 cm/V, a current of 1.44 mA, producing a range of measured average drift velocity between −77.1 and −55.8 μm/sec. Most prominent ZP peaks ranged from −60 mV to −52 mV, with most uniform and monodisperse plots recorded near the data set average at −52.73±4.57 mV. A low background electrolyte of phosphate solution conductivity of 245 μS with specific solution conductance of 185 μS/cm produced a high sample count rate at 839 kcps. CPMV T184C mutant produced a bimodal plot for several runs with signal peaks at 0 mV and −43.5 mV, with mobilities ranging from −3.48 and −3.35 mobility units at nonzero maxima, corresponding to an average drift velocity of −54.0 to −52.0 μm/s using an applied E-field of 15.53 V/cm, and a current of 1.62 mA. Some clustering due to limited disulfide interaction may have occurred, yielding very slow measured migration velocities, producing near zero mobilities and corresponding zeta potentials. Represented plot at average nonzero peak value of −43.15±4.74 mV with a reasonable sample count rate of 78 kcps at low solution conductance measured at 279 μS with specific conductance of 131 μS/cm. CPMV L189C displayed the most uniform mobility plots ranging from −3.75 to −2.89 units, corresponding to −53.7 to −41.4 μm/s at an applied voltage of 14.32 V/cm and 1.33 mA. Zeta potentials produced consistent peaks within −46.35 to −37 mV with an average run of −42.14 mV with a half-width of 4.97 mV. Very high sample count rate of 998 kcps was observed with a low solution conductance of 225 μS and specific conductance of 100 μS/cm

potential values using Smoluchowski approximation according to electrokinetic theory of colloids at solution conditions.

In order to determine the double-layer thickness, the ionic strength of solution was calculated for N dissociating ions using the combination of 1:1 and 1:2 electrolytes:

$$I_c = \frac{1}{2}\sum_{i=1}^{N} c_i z_i^2$$

$$KH_2PO_4 \rightarrow K^+ + H2PO_4^-$$
$$K_2HPO_4 \rightarrow 2K^+ + HPO_4^{-2}$$

At a pH of 7.0 of stock virus solutions and pK_a of 7.21 for monobasic potassium phosphate, proton dissociation is calculated by:

$$\frac{[HPO_4^{-2}]}{[H_2PO_4^-]} = \frac{mol\,DB}{mol\,MB} = \frac{F_{DB}}{F_{MB}} = 10^{pH-pKa} = 0.616$$

$$I_c = \frac{1}{2}\{[K^+](+1)^2 + [H_2PO_4^{-1}](-1)^2 + (2)[K^+](+1)^2[HPO_4^{-2}](-2)^2\}$$
$$= (0.5)F_{MB}C_T + F_{MB}C_T + 2F_{DB}C_T + F_{DB}C_T \quad (4)$$

where C_T is the total buffer concentration of final solution and F_{MB} and F_{DB} are mole fractions of monobasic and dibasic ions, respectively. The ionic strength is 2.34×10^{-3}M for CPMV wild type, and 1.406×10^{-3}M for CPMV 184c and 189c mutants, where the ionization of water is neglected.

Using the permittivity of vacuum ε_0 (8.854542×10^{-12}F m^{-1}, the relative permittivity ε_r of water (25°C) of 78.54, producing a liquid permittivity of $\varepsilon = \varepsilon_r \varepsilon_0 = 6.955 \times 10^{-10}$ F m^{-1}, Boltzmann's constant k_B=1.3807×10^{-23}JK^{-1}, temperature of 298.15 K, electronic charge e of 1.6022×10^{-19}C, ionic strength of solution I_c calculated in mol/l, and avogadro constant N_{avo} of 6.022×10^{23}mol^{-1}, the double-layer thickness κ^{-1} is determined (Hunter 1981).

$$\kappa^{-1} = \{(\varepsilon k_B T)/(2e^{2*}1000^*/I_c N_{avo})\}^{0.5}$$

κ^{-1} is calculated at 6.29 nm for wild type CPMV and 8.11 nm for mutants 184c and 189c, and represents the range of repulsive forces between each particle. Using a reported diameter of 32 nm for CPMV particles[6], the dimensionless κa values are 2.54 for CPMV-WT and 1.97 for CPMV mutants, suggesting that Henry's equations[34] is more applicable for most precise $f(\kappa a, \zeta)$ correlations between electrophoretic mobility and zeta potential.

Observed trends in peak zeta potential values (Fig. 4) between CPMV-WT, T184C, and L189C can be explained by the extent of water solvability to alter the local surface ion concentration, partial dissociation of cysteine to form a thiolate

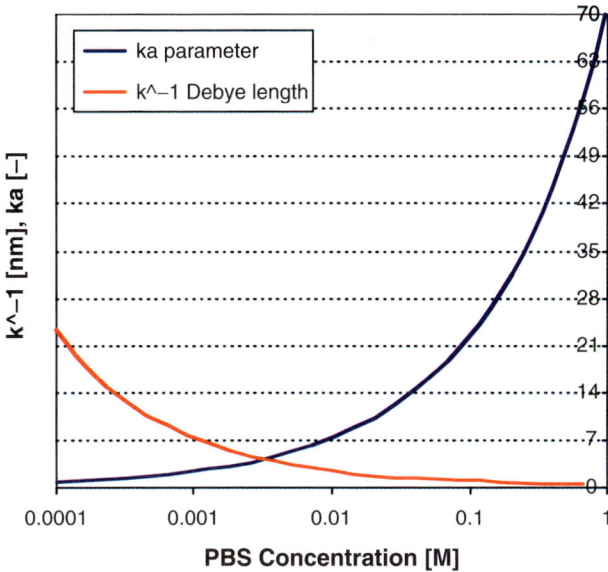

Fig. 4 Calculated electrokinetic radius and Debye parameter as a function of total buffer concentration. Debye length κ^{-1} is calculated as a function of total phosphate concentration for virus samples (*red*), based on dependence of ionic strength and relation between ionic strength and total concentration at solution pH, with abscissa values in logarithmic scale. All virus solutions exhibit the same plot because of similar dependence on solution ionic strength to calculate double-layer thickness. At greater concentration, double-layer thickness rapidly decreases until attractive forces between virions dominate, leading to mass aggregation effects, which would occur below 0.001 M to produce a double-layer thickness greater of approximately 5 nm. Electrokinetic radius κa is plotted as a function of total buffer concentration for CPMV virus samples. Electrophoretic radii as a function of buffer concentration for CPMV viruses is plotted in *purple*. A combination of 1:1 and 1:2 monobasic and dibasic potassium phosphate ions comprise the buffer, whose total phosphate concentration is inverse square root dependent on the electrophoretic radii. At concentrations beyond 0.5 M, Smoluchowski approximation for determination of zeta potential is most applicable with minimal error. At a constant geometric radius of 16 nm, only the solution ionic strength dependent on total PBS concentration; therefore all virus samples would exhibit the same dependency on κa

state to attract surface cations, and dynamic sulfohydryl–disulfide displacement reactions between neighboring cysteine groups. Using a pK_a of 8.3 for cysteine sulfohydryl groups, the percent dissociation to thiolate anions at pH 7.2 is 7.32%, suggesting that an additional 60 surface exposed cysteines introduced by point mutations at each subunit coat protein would accumulate more solution cations and increase the electrostatic potential. Substitution of leucine by cysteine at position 189 enhances solvation by water, allowing a greater density of ions to accumulate on the surface, providing a lesser negative increase in surface potential by accumulation of more cations at the Stern layer from potassium phosphate buffer. Despite the monodispersity of the samples used, a fairly wide range of zeta potentials may be a result of dynamic sulfohydryl–disulfide exchange reactions which fundamentally alter the surface charge density and mobilities. The broader profiles for T184C may be attributed to wider separation between cysteines at each fivefold axis compared with L189C, causing more pronounced disulfide aggregation, leading to greater polydispersity, more variable migration velocities, and broader corresponding zeta potentials. From wild type, small subunit 184 threonine's hydroxyl group is capable of hydrogen bonding with water, unlike nonpolar leucine at 189, implying that cysteine substitution at leucine would most likely contribute to greater local ion recruitment at the virus surface. Furthermore, by inspection of the small subunit coat protein molecular structure, the presence of neighboring cysteine at residue 14 is closer to L184 than T189, suggesting that intraviral disulfide formation is more probable with T184C than L189C. Therefore, the cysteine thiolate state is more favorable for L189C, allowing greater accumulation of surface cations, achieving a more positive electrostatic potential.

A general issue for colloidal solutions is the requirement of higher salt concentration required to satisfy Smoluchowski for κa>>1 (Fig. 3), but a very low salt concentration is ideal to achieve a large double-layer thickness κ^{-1} to avoid aggregation effects. In order to anticipate changes in electrostatic potential and confirm applicability of electrokinetic equations, dependence on zeta potential with modified solution conditions can be obtained for total buffer concentration. By relating ionic strength I_C with total buffer concentration C_T, pK_a, and solution pH, a dependence on total concentration at solution conditions on electrokinetic radius and Debye length is plotted in Fig. 3. Using the definition of ionic strength for our electrolyte system, and acid base equilibrium relations, κ^{-1} is found to vary with total salt concentration by

$$69.811\sqrt{C_T}, \text{ while } I_C = C_T \frac{\left[1+3\left(10^{pH-pKa}\right)\right]}{\left(1-10^{pH-pKa}\right)}.$$

At more basic conditions, a greater number of potassium ions are dissociated per mole of total buffer concentration, attracting more cations on the surface and altering the particle surface charge density.

References

Carraway KL, Triplett RB (1970) Reaction of carbodiimides with sulfhydryl protein groups. Biochim Biophys Acta 944:297–307

Chen J, Saeki F, Wiley BJ et al (2005) Gold nanocages: bioconjugation and their potential use as optical imaging contrast agents. Nano Lett 5:473–477

Douglas T, Young M (1999) Virus particles as templates for material synthesis. Adv Mater 11:679–681

Duncan R (2003) The dawning era of polymer therapeutics. Nat Drug Discov 2:347–360

Fisher AJ, Johnson JE (1993) Ordered RNA controls capsid architecture in an icosahedral animal virus. Nature 361:176–179

Fisher AJ, McKinney BR, Wery JP, Johnson JE (1992) Crystallization and preliminary data analysis of flock house virus. Acta Cryst B48:515–520

Gopidas KR, Whitesell JK, Fox MA (2003) Nanoparticle-cored dendrimers: synthesis and characterization. J Am Chem Soc 125:6491–6502
Hainfeld JF, Furuya FR (1992) A 1.4-nm gold cluster covalently attached to antibodies improves immunolabeling. J Histochem Cytochem 40:177–184
Hermanson GT (1996) Bioconjugate techniques. Academic, San Diego
Hückel E (1924) Die Kataphorese der Kugel. Phys Z 25:204–210
Hunter RJ (1981) Zeta potential in colloid science. Principles and applications. Academic, San Diego
Khorana HG (1993) The chemistry of carbodiimides. Chem Rev 53:145
Kim S, Lim YT, Soltesz EG et al (2004) Near-infrared fluorescent type II quantum dots for sentinel lymph node mapping. Nat Biotechnol 22:93–97
Krishnaswami S (2003) Hybrid virus-polymer materials. 1. Synthesis and properties of PEG-decorated cowpea mosaic virus. Biomacromolecules 4:472–476
Lin T, Porta C, Lomonossoff G (1996) Structure-based design of peptide presentation on a viral surface: the crystal structure of plant/animal virus chimera at 2.8A resolution. Fold Des 1:179–187
Maeda H, Sawa T, Konno T (2001) Mechanism of tumor-targeted delivery of macromolecular drugs, including the EPR effect in solid tumor and clinical overview of the prototype polymeric drug SMANCS. J Control Release 74:47–61
Michalet X, Pineaud FF, Bentolila LA et al (2005) Quantum dots for live cells, in vivo imaging, diagnostics. Science 307:538–544
Oldenburg SJ, Jackson JB, Westcott SL, Halas NJ (1999) Infrared extinction properties of gold nanoshells. Appl Phys Lett 75:2897–2899
O'Neal DP, Hirsch LR, Halas NJ et al (2004) Photo-thermal tumor ablation in mice using near infrared-absorbing nanopartricles. Cancer Lett 209:171–176
Parak WJ, Pelligrino T, Micheel CM et al (2003) Conformation of oligonucleotides attached to gold nanocrystals probed by gel electrophoresis. Nanoletters 3:33–36 2003
Portney NG, Singh K, Chaudhary S et al (2005) Organic and inorganic nanoparticle hybrids. Langmuir 21:2098–2103
Rojas-Chapana JA, Correa-Duarte M, et al (2004) Enhanced introduction of gold nanoparticles into vital Acidothiobacillus ferrooxidans by carbon nanotube-based microwave electroporation. Nano Lett 4:985–988
Shenton W, Douglas T, Young M et al (1999) Inorganic-organic nanotube composites from template mineralization of tobacco mosaic virus. Adv Mater 11:253–256
Smoluchoswki M (1903) Bull Acad Sci Cracovie 182–199
Stern O (1924) Zur theorie der elektrolytischen doppelschicht. Z Elektrochem 30:508–516
Strable E, Johnson JE, Finn MG (2004) Natural nanochemical building blocks: icosahedral virus particles organized by attached oligonucleotides. Nanoletters 4:1385–1389
Tshikhudo TR, Wang Z, Brust M (2004) Biocompatible gold nanoparticles. Mater Sci Technol 20:980–984
Wang Q, Kaltgrad E, Lin T et al (2002a) Natural supramolecular building blocks: wild-type cowpea mosaic virus. Chem Biol 9:805–811
Wang Q, Lin T, Johnson JE, Finn MG (2002b) Natural Supramolecular Building Blocks: Cysteine-Added Mutants of Cowpea Mosaic Virus. Chem Biol 9:813–819
Wang Q, Lin T, Tang L, Johnson JE, Finn MG (2002c) Icosahedral virus particles as addressable nanoscale building blocks. Angew Chem Ing Ed 41:459–462
Wong N, Kam S, Jessop TC, Wender PA Dai H (2004) Nanotube molecular transporters: internalization of carbon nanotube-protein conjugates into mammalian cells. J Am Chem Soc 126:6850–6851
Worden JE, Shaffer AW, Huo Q (2004) Controlled functionalization of gold nanoparticles through a solid phase synthesis approach. Chem Commun 518–519

A Library of Protein Cage Architectures as Nanomaterials

M.L. Flenniken (✉), M. Uchida, L.O. Liepold, S. Kang,
M.J. Young, T. Douglas

Contents

Introduction	72
Protein Cage Discovery from Hyperthermophilic Archaea	74
Protein Cages from Mesophilic Organisms	77
Protein Cages for Inorganic Nanoparticle Synthesis	77
Protein Cages for Medical Imaging	82
Protein Cages for Targeted Therapeutic and Imaging Agent Delivery	83
Asymmetric Derivatization of Inherently Symmetric Protein Cage Architectures	86
In Vivo Study of Protein Cage-Mediated Materials for Medical Applications	87
Introduction of Multiple Functionalities on a Single Protein Cage Architecture	87
Conclusion	88

Abstract Virus capsids and other structurally related cage-like proteins such as ferritins, dps, and heat shock proteins have three distinct surfaces (inside, outside, interface) that can be exploited to generate nanomaterials with multiple functionality by design. Protein cages are biological in origin and each cage exhibits extremely homogeneous size distribution. This homogeneity can be used to attain a high degree of homogeneity of the templated material and its associated property. A series of protein cages exhibiting diversity in size, functionality, and chemical and thermal stabilities can be utilized for materials synthesis under a variety of conditions. Since synthetic approaches to materials science often use harsh temperature and pH, it is an advantage to utilize protein cages from extreme environments. In this chapter, we review recent studies on discovering novel protein cages from harsh natural environments such as the acidic thermal hot springs at Yellowstone National Park (YNP) and on utilizing protein cages as nano-scale platforms for developing nanomaterials with wide range of applications from electronics to biomedicine.

M.L. Flenniken
University of California, San Francisco, Microbiology and Immunology Department, 600 16th Street, Genentech Hall S576, Box 2280, San Francisco, CA 94158–2517, USA
e-mail: michelle.flenniken@ucsf.edu

Abbreviations CCMV Cowpea chlorotic mottle virus, Dps DNA-binding proteins from starved cells, Dps-L Dps-like protein, Hsp Heat shock protein, MRI Magnetic resonance imaging, STIV *Sulfulobus* turreted icosahedral virus, TMV Tobacco Mosaic Virus, YNP Yellowstone National Park

Introduction

In nature, proteins orchestrate the formation of elaborate inorganic structures, for example, the single-celled algae, *Emiliania huxleyi*, form intricate calcium carbonate ($CaCO_3$) structures called coccoliths (Fig. 1a) (Wikipedia 2005). In comparison, synthetic preparations of $CaCO_3$ result in a far more limited range of morphologies. Coccolith formation is controlled by proteins that direct the assembly of crystallites into intricate 3D assemblies. The degree of control exhibited by this and other natural systems is an inspiration for materials scientists.

Viral capsids are also naturally occurring, intricate assemblies that serve to house, protect, and deliver nucleic acid genomes to specific host cells. Therefore, their structures must be robust enough to survive diverse conditions, yet sufficiently dynamic to release their genome into host cells. Proteins are the building blocks of viral capsids; therefore, protein–protein interactions dictate their 3D structure. Typically, protein motifs on the interior are involved in packaging nucleic acid, whereas those on the exterior are involved with cell recognition and attachment. Viral capsids devoid of their nucleic acid genomes can be thought of as nanocontainers. The diversity of these nanocontainers is seemingly endless, since viruses are ubiquitous with life. In our investigations of archaeal viruses found in the acidic hot (>90°C) springs of Yellowstone National Park (YNP), we discovered and structurally characterized the *Sulfolobus* turreted icosahedral virus (STIV), which presents elaborate turret-like structures on its exterior (Rice et al. 2001, 2004; Synder et al. 2003) (Fig. 1b).

Both the coccolith and archaeal virus are examples of naturally occurring 3D assemblies whose architecture is dictated by proteins. Inspired by nature, we have selected a bio-mimetic approach to nanomaterials synthesis that is bioassisted. We utilize protein cage architectures to serve as size-constrained reaction vessels and chemical building blocks (Douglas and Young 1998, 1999; Douglas et al. 2002b, 2004; Flenniken et al. 2004; Kelm et al. 2005b).

Protein cage architectures, 10–100 nm in diameter, are self-assembled hollow spheres derived from viruses and other biological cages, including heat shock proteins (Hsp), DNA-binding proteins from starved cells (Dps), and ferritins. These architectures play critical biological roles. For example, heat shock proteins are thought to act as chaperones that prevent protein denaturation, and ferritins are known to store iron (which is both essential and toxic) as a nanoparticle of iron oxide (Harrison and Arosio 1996; Narberhaus 2002). While each of these structures has evolved to perform a unique natural function, they are similar in that they are all essentially proteinaceous containers with three distinct surfaces (interior, exterior, and subunit interface) to which one can impart function by design. Protein cage architectures have demonstrated

A Library of Protein Cage Architectures as Nanomaterials 73

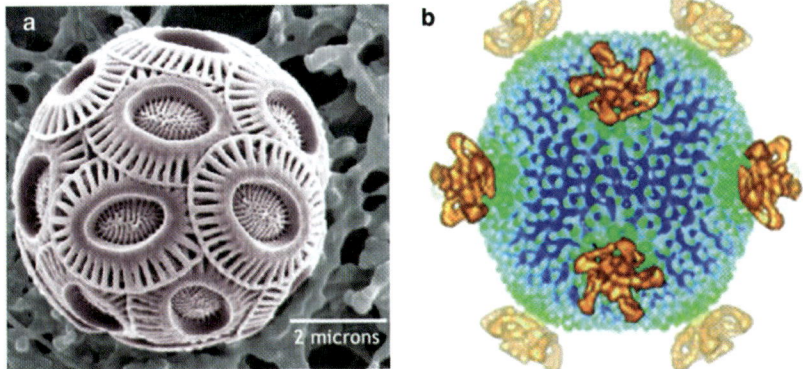

Fig. 1a,b Biological assemblies. **a** *Emiliania huxleyi* formed $CaCO_3$ coccolith structures arranged in a coccosphere, ~6 mM in diameter. (Used with permission from the Wikipedia:Free Encyclopedia (http://en.wikipedia.org, with permission; picture by Dr. Markus Geisen) **b** Cryo-TEM image reconstruction of the ~74-nm-diameter STIV capsid with turret-like projections extending from each of the fivefold vertices. (Rice et al. 2004, with permission)

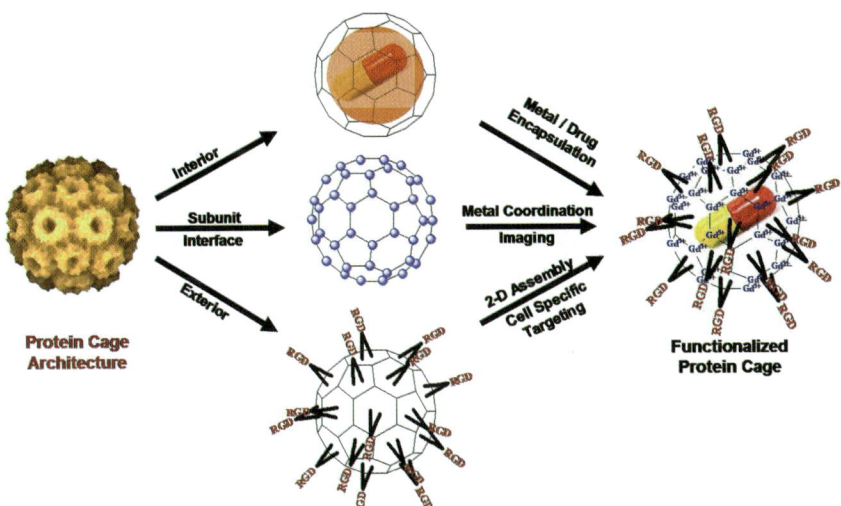

Fig. 2 Schematic representation of protein cage functionalization. Protein cage architectures have three surfaces (interior, subunit interface, and exterior) amenable to both genetic and chemical modification. (Mark Allen)

utility in nanotechnology with applications including inorganic nanoparticle synthesis and the development of targeted therapeutic and imaging delivery agents (Allen et al. 2002, 2003, 2005; Bulte et al. 1994a, 1994b, 2001; Chatterji et al. 2002, 2004a, 2004b; Douglas and Stark 2000; Flenniken et al. 2005, 2006; Gillitzer et al. 2002; Hikono et al. 2006; Liepold et al. 2007; Raja et al. 2003a, 2003b; Schlick et al. 2005; Wang et al. 2002a, 2002b, 2002c) (Fig. 2).

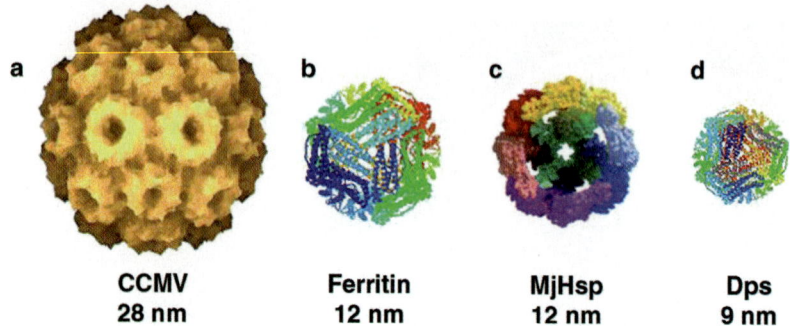

Fig. 3a–d Protein cage library. **a** Cowpea chlorotic mottle viral (CCMV) capsid cryo-image reconstruction. (Reddy et al. 2001, with permission from VIPER). **b** Ribbon diagram of human H-chain ferritin; our library also includes horse spleen and *Pyrococcus furiosus* ferritins. **c** Small heat shock protein (Hsp) from *Methanococcus jannaschii*. (Kim et al. 1998a, with permission). **d** Ribbon diagram of Dps (DNA binding protein from starved cells) protein, our library includes a Dps protein from *Listeria innocua* and Dps-like proteins from *Sulfolobus solfataricus* and *Pyrococcus furiosus*

Protein cage architectures are naturally diverse; each has unique attributes (including size, structure, solvent accessibility, chemical and temperature stability, structural plasticity, assembly and disassembly parameters, and electrostatics) useful to particular applications. Importantly, one can capitalize on these features or alter them via genetic or chemical modification. Atomic level structural information identifies the precise location of amino acids within protein cage architectures and in turn allows for the rational inclusion, exclusion, and substitution of amino acid(s) (at the genetic level) resulting in protein cages with novel functional properties.

In this work, we review recent strides our lab has taken toward the development of protein cage architectures as nanomaterials for bioengineering and biomedicine. We have developed a library of protein cages (including Cowpea chlorotic mottle virus [CCMV], ferritin, Hsp, and Dps) as size-constrained reaction vessels and as platforms for genetic and chemical modification (Fig. 3). This protein cage library consists of a small sampling of the great diversity of naturally available protein cage architectures. Utilization of protein cages for applications in nanotechnology often requires their use in harsh synthetic conditions. Therefore, we have focused some of our efforts on novel protein cage discovery from harsh natural environments.

Protein Cage Discovery from Hyperthermophilic Archaea

The quest to find new, potentially stable protein cages has led our group to the hot springs of Yellowstone National Park. Within these extreme environments (pH 1.5–5.5; temperature range of 70°C to >100°C), organisms in the archaeal domain such as *Sulfolobus solfataricus* are well represented (Rice et al. 2004). *Sulfolobus* and other hyperthermophiles serve as hosts for novel and largely uncharacterized viruses. Of the approximately 5,100 known viruses, only 36 archaeal viruses, or virus-like particles, have been described to date (International Committee on Taxonomy of Viruses, http://www.ncbi.nlm.nih.gov/ICTVdb/Ictv/index.htm (Ackerman 2001). Prior to our investigation, no viruses of *Sulfolobus* from YNP had been described (Rice et al. 2004). In 2001, we reported six unique particle morphologies isolated from YNP, three of which were similar to viruses isolated from Iceland or Japan and the other three exhibited novel morphologies (Rice et al. 2001) (Fig. 4). One YNP virus, named

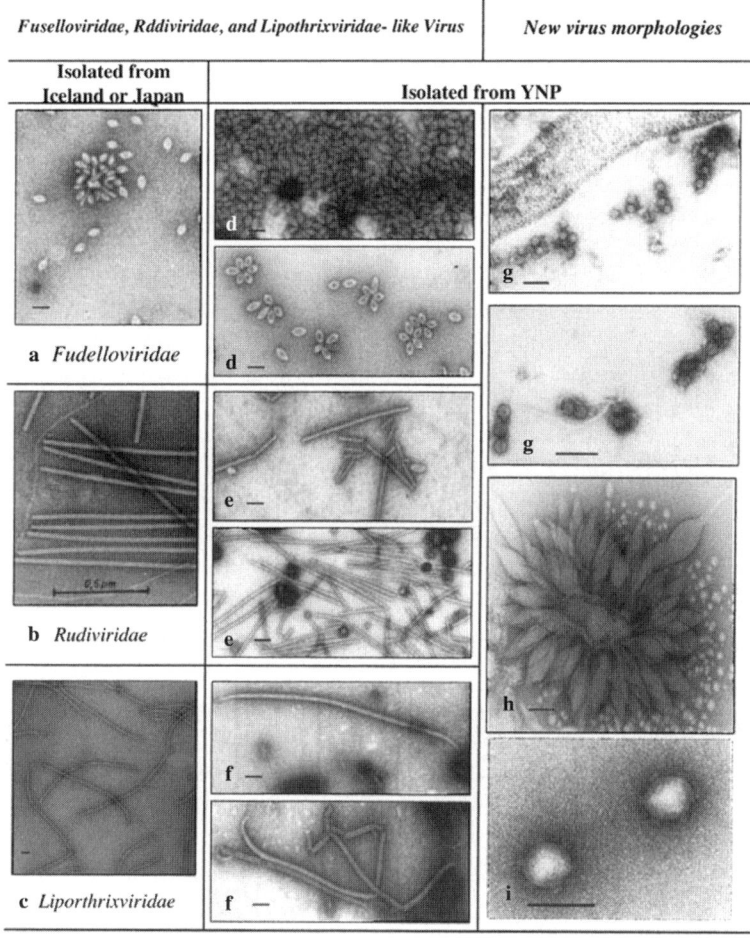

Fig. 4a–i Transmission electron micrographs of virus and virus-like particles from YNP. **a** SSV1 Fusellovirus, **b** SIRV Rudivirus, and **c** SIFV Lipothrixvirus previously isolated from thermal areas of Japan or Iceland (courtesy of W. Zillig, Max-Planck Institut für Biochemie, Martinsried, Germany). **d** SSV-like, **e** SIRV-like, and **f** SIFV-like particle morphologies isolated from YNP thermal features. **g–i** Virus-like P articles isolated from YNP thermal features. Bars indicate 100 nm. (Rice et al. 2001, with permission)

STIV, has an approximately 74-nm diameter capsid from which turret-like projections extend (Rice et al. 2004) (Fig. 1b). The pseudo-T=31 structure of STIV is unique. A structural comparison of STIV, bacterial phage PRD1, and human adenovirus revealed tertiary and quaternary structural similarities suggestive of a common ancestry that predates the division of life into three domains (Eukarya, Bacteria, and Archaea) more than 3 billion years ago (Rice et al. 2004).

In addition to new viruses, hyperthermophilic organisms are hosts to other protein cage architectures, including Hsp cages, ferritins, Dps and Dps-like proteins. Genome sequencing of many archaea since the first, *Methanococcus jannaschii* (isolated from a 2,600-μ-deep hydrothermal vent in the Pacific Ocean) in 1996 has aided the discovery and study of protein cages (Bulte et al. 1996). The small heat shock protein from *M. jannaschii* (*Mj*Hsp) assembles into a 12-nm-diameter cage with octahedral symmetry (Kim et al. 1998a, 1998b) (Fig. 3). The crystal structure of this hollow spherical complex determined that there are eight triangular and six square pores that allow free exchange between the interior and bulk solution, a useful feature for applications in nanotechnology, including loading therapeutics into protein cage architectures (Flenniken et al. 2003, 2005; Kim et al. 1998a).

Utilizing genomic information, our lab recently identified Dps-like protein cages from *Sulfolobus solfataricus* (*Ss*Dps-L) and *Pyrococcus furiosus* (*Pf*Dps-L) (Maeder et al; 1999; Ramsay et al. 2006; Wiedenheft et al. 2005). Characterization of the *Ss*Dps-L determined that it serves to protect *S. solfataricus* against oxidative stress and in the process mineralizes a nanoparticle of iron oxide within its cage structure (Ramsay et al. 2006; Wiedenheft et al. 2005). The *Ss*Dps-L protein was identified based on predicted secondary and tertiary structural similarity to a Dps protein from *Listeria innocua*, while the *Pf*Dps-L has a very high sequence homology to the *Ss*Dps-L (Su et al. 2005; Wiedenheft et al. 2005). These 9.7-nm-diameter protein cages self-assemble from 12 protein monomers and represent the smallest of the protein cages in our library (Ramsay et al. 2006; Wiedenheft et al. 2005) (Fig. 5).

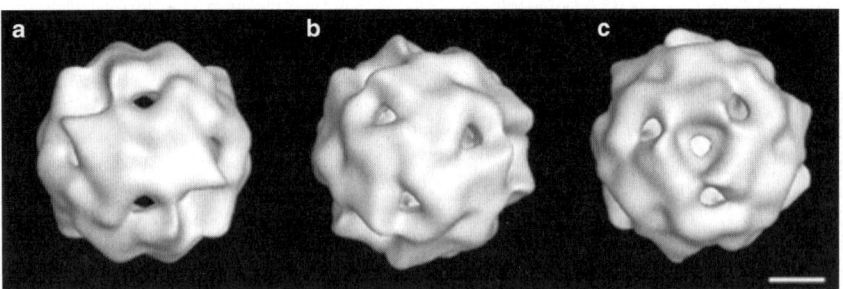

Fig. 5a–c 3D image reconstruction of the assembled *Ss*Dps-L cage. Surface-shaded views of reconstructed negative-stained images displayed along the (**a**) twofold (2F) axis, and (**b** and **c**) along the two nonequivalent environments at each end of the threefold (3F) axis. Scale bar: 2.5 nm. (Wiedenheft et al. 2005, with permission)

Protein Cages from Mesophilic Organisms

Protein cages isolated from thermophilic environments are desirable as building blocks for nanotechnology due to their potential stability in harsh reaction conditions including high temperature and pH extremes. Interestingly, one of the most stable protein cage architectures, ferritin, is commonly found in mesophilic organisms, including animals, plants, and microbes. For example, horse spleen ferritin exhibits broad pH (pH 2–8) and temperature stability (<70°C). Ferritins are involved in iron sequestration, which they accomplish through the oxidation of soluble Fe(II) using O_2 (Chasteen and Harrison 1999; Harrison and Arosio 1996). This oxidation results in the formation of a nanoparticle of Fe_2O_3 encapsulated (and rendered nontoxic) within the protein cage. High charge density on the inner surface of the protein cage promotes this reaction, which is assisted by an enzymatic (ferroxidase) activity in some ferritin subunits (Chasteen and Harrison 1999; Douglas and Riipoll 1998; Harrison and Arosio 1996). Ferritins are made up of 24 subunits, which form a spherical cage 12 nm in diameter (Harrison et al. 1976; Harrison and Arosio 1996). The ferritin family also includes the 24 subunit bacterioferritins and the Dps class of proteins, which assemble from 12 monomers. We, and others, have employed ferritin and Dps cage architectures for mineralization of inorganic nanoparticles (Allen et al. 2002, 2003; Douglas 1996; Douglas and Stark 2000; Ensign et al. 2004; Hosein et al. 2004; Wiedenheft et al. 2005). Recently, we utilized a genetically modified recombinantly expressed human H-chain ferritin as a platform for cell-specific delivery of imaging agents in vitro (Uchida et al. 2006).

Mesophilic viral capsids have also exhibited utility for applications in nanotechnology (Allen et al. 2005; Chatterji et al. 2002, 2004a, 2004b, 2005; Douglas and Young 1998, 1999; Douglas et al. 2002a; Flenniken et al. 2004; Gillitzer et al. 2002; Klem et al. 2003; Liepold et al. 2007; Mao et al. 2003, 2004; Raja et al. 2003a, 2003b; Schlick et al. 2005; Strable et al. 2004; Wang et al. 2002a, 2002b, 2002c). We have routinely employed the CCMV capsid as a scaffold for chemical conjugation of biologically important molecules, including imaging agents (fluorescein, Texas red, gadolinium chelators), and as a size-constrained reaction vessel for mineralization of metallic and metal oxide nanoparticles (Douglas and Young 1998; Gillitzer et al. 2002; Klem et al. 2003; Liepold et al. 2007). The CCMV capsid assembles from 180 copies of a single protein into a 28-nm icosohedral shell (Johnson and Speir 1997; Speir et al. 1995; Shao et al. 1995). Wild type and genetically engineered CCMV capsids are purified from both the natural cowpea plant host and a yeast heterologous expression system (Brumfield et al. 2004).

Protein Cages for Inorganic Nanoparticle Synthesis

In our initial work, CCMV capsids were employed for the mineralization of polyoxometalate species (paratungstate and decavanadate) (Douglas and Young 1998). The virion's interior cavity constrains mineral growth, resulting in a spherical nanoparticle with a maximum diameter of approximately 24 nm (Douglas and Young 1998). The exterior and interior of the CCMV capsid are chemically distinct environments. The interior surface is more positively charged than the exterior, thus allowing it to serve as a nucleation site for aggregation of anionic precursors and crystal growth. After nucleation, the size and shape of the mineral are defined by the interior of the virion. In addition to the endogenous properties of size, shape, and delineation of charge on the interior and exterior surfaces, the CCMV viral capsid undergoes a pH-dependant structural transition (gating) (Schneemann and Youg 2003; Speir et al. 1995). This pH-dependant reversible gating or swelling is a structural change of the virus capsid that results in a 10% increase of viral dimension at pH greater than 6.5 in the absence of metal cations (at pH <6.5 the virion is in its closed confirmation) (Speir et al. 1995). During the polyoxometalate mineralization reaction, the pH was lowered entrapping the mineral within the viral capsid (Douglas and Young 1998).

In addition to utilizing the inherent properties of the CCMV protein cage as described above, we can dramatically change these properties via genetic engineering. Specifically, site-directed polymerase chain reaction-mediated mutagenesis is employed to alter the DNA encoding the viral capsid protein. Protein expression results in the replacement of wild type amino acids and functional groups with novel functional groups at precisely defined locations due to the quaternary structure of the virion. Additionally, genetic modifications allow for the introduction of novel peptide sequences as extensions from the N- and C-termini or within the loop regions of the subunit structure (Liepold et al. 1983).

Inspired by the iron storage protein cage, ferritin, and its ability to mineralize iron oxide nanoparticles in a size-constrained fashion, we constructed the subE mutant of CCMV. SubE was made by replacing eight of the positively charged amino acid residues (5-Arg, 3-Lys) on the N-terminus of the protein with negatively charged glutamic acids (Glu). This created a cage with a negatively charged interior, illustrating the plasticity of the CCMV cage toward genetic manipulation (Brumfield et al. 2004; Douglas et al. 2002b). The electrostatically altered viral protein cages did not differ in overall structure from wild type CCMV, but they exhibited different mineralization capabilities (Brumfield et al. 2004). SubE's negatively charged interior promoted interaction with cationic Fe(II) ions and subsequent oxidative hydrolysis led to the formation of approximately 24-nm particles of lepidocrocite (γ-FeOOH) constrained within the interior cavity of the protein cage (Douglas et al. 2002b).

We have developed the library of protein cages available for biomineralization and chemical derivatization to include many protein cage architectures of both viral and nonviral origin. Protein cage architectures housing iron oxide (magnetite) have potential to serve in magnetic memory storage and as contrast agents for magnetic resonance imaging (MRI). In addition to CCMV and ferritin, we have mineralized iron oxide nanoparticles within a number of additional protein cages. These include the Dps protein cage from *Listeria innocua* (*Li*Dps), the *Ss*Dps-L cage from *S. solfataricus*, the small Hsp (*Mj*Hsp) cage from *M. jannaschii*, the ferritin cage from horse spleen, human H-chain ferritin and the ferritin from the hyperthermophile

Pyrococcus furiosus (Allen et al. 2002, 2003; Douglas and Mann 1995; Flenniken et al. 2003; Klem et al. 2005b; Parker et al. 2008; Uchida et al. 2006; Wiedenheft et al. 2005). The *Li*Dps was shown by others to mineralize Fe as a nanoparticle of an as-yet unidentified ferric oxyhydroxide (Stefanini et al. 1999; Yang et al. 2000). We utilized the *Li*Dps cage as a size-constrained reaction vessel for Fe_3O_4 mineralization under the nonphysiologic conditions of elevated pH and temperature (pH 8.5, 65°C) and in the presence of substoichiometric amounts of oxidant (H_2O_2) (Allen et al. 2002). Under these conditions, with 400 Fe(II) per cage, ferrimagnetic nanoparticles of Fe_3O_4 (4 nm in diameter) formed inside the *Li*Dps cage (Allen et al. 2002) (Fig. 6).

Analogous to ferritin and *Li*Dps protein cages, the *Mj*Hsp cage has been shown to act as a size-constrained reaction vessel for the formation of iron oxide (ferrihydrite). Transmission electron microscope imaging of the 9-nm cores within Hsp cages, as compared to iron oxide formation in the absence of protein cages, underscore the importance of the protein cage to control and constrain mineral growth (Flenniken et al. 2003) (Fig. 7). Like iron oxide formation in the CCMV SubE protein cage, mineralization within the Hsp cage illustrates the ability of many protein cage platforms to function similarly as nanocontainers for biomineralization.

Based on the structural similarity of *Ss*Dps-L to the Dps protein from *L. innocua*, we explored the in vitro mineralization capability of the 10-nm-diameter *Ss*Dps-L protein cage (Wiedenheft et al. 2005). This analysis revealed that unlike ferritins, *Ss*Dps-L mineralizes iron oxide more efficiently in the presence of H_2O_2 as compared to ferritins that catalyze Fe(II) oxidation with O_2. Ferritin protein cages are thought to serve as iron storage proteins, whereas Dps and Dps-like proteins mineralize iron as a consequence of hydrogen peroxide reduction, therefore serving as antioxidants that protect the organism during oxidative stress (Wiedenheft et al. 2005).

In addition to polyoxometalates and iron oxides, we and others have expanded the range of inorganic nanoparticles formed within these architectures to include important semiconducting materials (Klem et al. 2005b). Ferritin architectures have

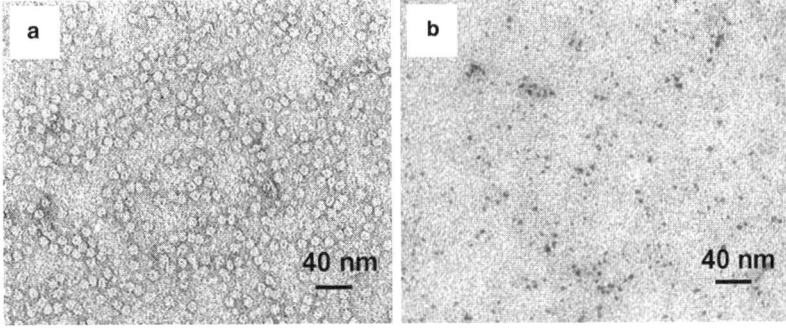

Fig. 6a, b Transmission electron micrographs of mineralized *L. innocua* Dps cage. *Li*Dps mineralized with g-Fe2O3 (**a**) stained with uranyl acetate and (**b**) unstained. (Allen et al. 2002, with permission)

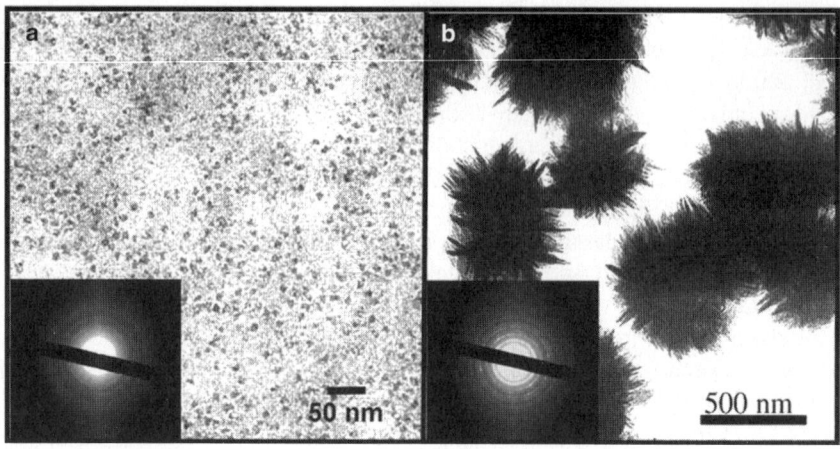

Fig. 7a, b Transmission electron micrographs of iron oxide mineralization. **a** Iron oxide cores inside Hsp cages; scale bar, 50 nm. **b** TEM of control sample illustrating the bulk precipitation of iron oxide in the absence of Hsp protein cages; scale bar, 500 nm. The insets show the electron diffraction from each sample. (Flenniken et al. 2003, with permssion)

been utilized as a constrained template for the synthesis of CoPt, FePt, ZnSe, Pd, CdS, Ag, and Eu_2O_3 nanoparticles (Allen et al. 2002, 2003; Douglas and Stark 2000; Douglas et al. 2002b; Klem et al. 2005a, 2005b, 2008; Varpness et al. 2005). *Li*Dps served as a size-constrained reaction vessel for the synthesis of two cobalt oxide minerals (Co_3O_4 and Co(O)OH) (Allen et al. 2003). In addition, utilizing the heat shock protein architecture, we have made CoPt and Pt nanoparticles (Klem et al. 2005a; Varpness et al. 2005). Inorganic nanoparticles templated by protein cages have been utilized in the fabrication of semiconducting devices. Yamashita and colleagues have developed what has been called the bionano process (BNP) for the fabrication of metal-oxide-semiconductor (MOS) such as a floating nanodot gate memory device or low-temperature polycrystalline silicon thin film transistor flash memory (Hikono et al. 2006; Ichikawa et al. 2007; Matsui et al. 2007; Miura et al. 2006; Yamada et al. 2007). The performance and characteristics of the MOS devices depend on the size, shape, and density of the nanodot array. They have recently demonstrated that the BNP allows control of these three parameters (Yamada et al. 2007).

Biomimetic approaches to materials synthesis have explored the interaction between proteins and minerals at their interface. Pioneering work by A. Belcher's group utilized the phage display library technique (originally developed to determine peptide–protein interactions for mapping antibody–antigen binding sites) to identify peptides that bound inorganic substrates normally not encountered by biological systems (Belcher et al. 2001; Burritt et al. 1996; Mao et al. 2004; Scott and Smith 1990; Seeman and Belcher 2002; Smith and Petrenko 1997; Whaley et al. 2000). In turn, when these peptides were present in mineralization reactions they directed the growth of specific crystal phases (Mao et al. 2004). The KTHEIHSPLLHK peptide specifically binds the $L1_0$ phase of CoPt (Mao et al. 2004). Genetic incorporation of this CoPt-binding peptide into the *Mj*Hsp cage interior (CP-Hsp) enabled phase-specific nucleation and size-constrained formation of CoPt nanoparticles with an average diameter of 6.5 nm (Klem et al. 2005a). The metallized CP-Hsp exhibited room temperature ferromagnetism, whereas HspG41C, a genetic variant of *Mj*Hsp with an internal cysteine, did not because of the nonspecific ordering of the CoPt mineral within its interior (Flenniken et al. 2003; Klem et al. 2005a). These results underscore the ability of specific peptide sequences to direct mineral formation. Protein cage architectures that incorporate these sequences (either inherently or because of genetic manipulation) are more than nanocontainers that serve as size-constrained reaction vessels. Their specific amino acids play a critical role in mineral formation in a manner that is not yet completely understood. By developing protein cage platforms for nanomaterial synthesis, we hope to gain insight into the mechanisms by which proteins direct mineral formation.

In addition to serving as size-constrained reaction vessels, protein cage architectures have been used as a scaffold to construct unique catalysts (Endo et al. 2007; Ensign et al. 2004; Kim et al. 2002; Miller et al. 2007; Nkere et al. 1997; Varpness et al. 2005). Native ferrihydrite-containing ferritins were utilized to catalyze the photoreduction of toxic Cr(VI) species to Cr(III) (Kim et al. 2002). In addition, ferrihydrite-containing ferritins were employed as catalysts for the photoreduction of Cu(II) to form colloidal Cu(0) nanoparticles ranging in size from 4 to 31 nm (Ensign et al. 2004). Treatment of iron and cobalt oxide containing ferritin protein cages with H_2 at elevated temperature resulted in conversion to the metallic Fe or Co nanoparticles without loss of morphology (Hosein et al. 2004). The catalytic activity of these highly reactive nanoparticles is being explored. Recently, TMV has been used as a scaffold to mimic light harvesting systems (Endo et al. 2007; Miller et al. 2007). Porphyrin complexes (Endo et al. 2007) or fluorescent chromophores (Miller et al. 2007) were attached in a spatially defined manner onto the inner surface of the TMV particles. Efficient energy transfer, from large numbers of donor chromophores to a single acceptor, has been demonstrated (Endo et al. 2007; Miller et al. 2007). For direct hydrogen production, Hsp cages were utilized as a template for Pt nanoparticle synthesis (Varpness et al. 2005). The Pt nanoparticles contained within Hsp exhibited comparable catalytic hydrogen production activity to hydrogenase enzymes (Varpness et al. 2005). Initial hydrogen production rates for this system were approximately tenfold greater than previously reported values from Pt colloids (Varpness et al. 2005).

The examples described thus far demonstrate the utility of protein cage architectures for inorganic nanomaterials synthesis and catalysis. In addition, metal-containing protein cages have potential medical applications. Protein cage architectures housing iron oxide (magnetite) nanoparticles have potential medical applications in imaging, serving as MRI contrast agents, and in cancer treatment by hyperthermia (Allen et al. 2002; Butle et al. 1994; Kawashita et al. 2005; Klem et al. 2005b; Luderer et al. 1983).

Protein Cages for Medical Imaging

We have developed protein cage platforms with fluorescence and MRI capabilities. Imaging capabilities were imparted via (1) magnetite mineralization within cage interiors, (2) utilization of endogenous sites to bind paramagnetic metal ions, (3) genetic incorporation of metal chelating peptides, and (4) chemical modifications, including covalently binding fluorophores and gadolinium chelators (Allen et al. 2002, 2005; Basu et al. 2003; Flenniken et al. 2003, 2005; Gillitzer et al. 2002; Liepold et al. 2007).

The road to developing the CCMV viral capsid as an MRI contrast agent has involved utilization of the endogenous properties of the cage, as well as both genetic and chemical modification. CCMV naturally coordinates divalent calcium (Ca^{2+}) ions at its quasi-threefold axis, but in vitro other metal ions, including paramagnetic gadolinium (Gd^{3+}), can bind these sites (Allen et al. 2005; Basu et al. 2003). The CCMV capsid is composed of 180 subunits, each of which binds a Gd^{3+} ion resulting in a particle with a high Gd^{3+} payload and in turn high relaxivity values (Allen et al. 2005). The T_1 and T_2 ionic relaxivities of water protons, as measured at 61 MHz, were 202 and 376 $mM^{-1}s^{-1}$, respectively; these are the highest values reported for a molecular paramagnetic material to date (Allen et al. 2005). Although Gd-CCMV exhibited high relaxivity values, the dissociation constant (K_d) for Gd^{3+} is 31 μM, insufficient for in vivo applications (Allen et al. 2005. In order to increase the binding affinity, we genetically incorporated the metal-binding motif from calmodulin into the CCMV architecture (Le Clainche et al. 2003; Liepold et al. 2007). At the same time, we explored the chemical attachment of GdDOTA onto endogenous lysines on CCMV. Although these two approaches both substantially increased the affinity for Gd^{3+}, there was a loss of relaxivity efficiency in both cases, as compared to Gd^{3+} bound to the endogenous binding sites of CCMV. The loss of relaxivity likely stems from two sources. First, in the case of the endogenous metal-binding sites of CCMV, there is more than one water molecule bound to Gd^{3+} (a higher number of Gd^{3+}-bound water molecules is preferred for efficient relaxivity properties) as compared to the genetically and chemically modified constructs. Secondly, the genetically and chemically modified constructs resulted in Gd^{3+} ions that are anchored to the protein capsid in a more flexible manner as compared to the endogenous binding site. These more flexible chelators negate much of the beneficial qualities of the slow rotational tumbling of the large protein cage structures.

Other studies chemically attached flexible Gd^{3+}-containing groups to the viral capsids; MS2, CPMV and Qβ (these capsids are similar in size to CCMV), which produced relaxivity values lower than the chemically modified CCMV-GdDOTA construct (Anderson et al. 2006; Prasuhn et al. 2007). Together these studies suggest that the relaxivity values improve with the use of shorter, more rigid Gd^{3+} linkers. Others have used hydroxypyridinone (HOPO)-based chelators to coordinate Gd^{3+}. In doing so, they took advantage of the higher number of ligand water molecules bound to Gd^{3+} and an ideal water exchange lifetime (Datta et al. 2008; Hooker et al. 2004). Also, their reaction scheme allowed for specific modification of the interior or the exterior of the MS2 viral capsid with GdHOPO. The results of this analysis produced relaxivity values slightly lower than that of the CCMV-GdDOTA construct but higher than the other Gd/capsid-based systems studied to date (Anderson et al. 2006; Prasuhn et al. 2007). The fit of their results to theoretical models provided insight into the factors that are responsible for producing relaxivity in Gd^{3+} based systems.

Protein Cages for Targeted Therapeutic and Imaging Agent Delivery

One of our goals is to develop protein cage architectures that serve as cell-specific therapeutic and imaging-agent delivery systems. Targeted therapeutic delivery systems can enhance the effective dose at the site, such as a tumor, while decreasing general exposure to the drug and its associated side effects (Allen and Cullis 2004). Protein cage architectures have three surfaces (interior, subunit interface, and exterior) amenable to both genetic and chemical modification. Figure 2 depicts a schematic representation of how each surface can play a distinct role in the development of new targeted therapeutic and imaging agent delivery systems. The cage interior can house therapeutics, the subunit interface incorporates gadolinium (an MRI contrast agent) and the exterior presents cell-specific targeting ligands (such as peptides and antibodies).

Protein cages have many beneficial attributes that are useful in their development as targeted therapeutic and imaging agent delivery systems. Their size falls into the nanometer range shown to localize in tumors due to the enhanced permeability and retention effect (Allen and Cullis 2004; Hashizume et al. 2000; Maeda et al. 2000). Their multivalent nature enables the incorporation of multiple functionalities (including targeting peptides and imaging agents) on a single protein cage. They are malleable to both chemical and genetic manipulation and can be produced in heterologous expression systems (including bacterial, yeast, and baculoviral systems) (Allen et al. 2002; Brumfield et al. 2004; Flenniken et al. 2003; Ramsay et al. 2006; Wiedenheft et al. 2005). In addition, detailed atomic resolution structural information enables the rational design of genetic mutants with specific functions, including cell-specific targeting (Douglas et al. 2002b; Flenniken et al. 2003, 2006; Gillitzer et al. 2002; Liepold et al. 2007; Uchida et al. 2006).

The CCMV protein cage was utilized in our first example of encapsulation of an organic compound (Douglas and Young 1998). The gating properties of the CCMV cage (swollen confirmation at pH 6.5 or higher, closed confirmation below pH 6.5) were utilized for the entrapment of an organic polyanion, polyanetholesulphonic acid (Douglas and Young 1998). We have also genetically introduced a redox-dependant chemical switch for reversible gating of CCMV and demonstrated in vitro functionality (C. Crowley et al., unpublished data). In this variant of CCMV, cysteines were introduced at the quasi-threefold axis, resulting in a capsid that is in a compact, closed conformation under oxidizing conditions and undergoes a transition to an open conformation under reducing conditions (C. Crowley et al., unpublished data).

In addition to the entrapment of materials on the interior, the CCMV protein cage and its genetic variants have been chemically derivatized on both the exterior and interior surfaces (M.L. Flenniken et al., unpublished data; Gillitzer et al. 2002; Suci et al. 2007) (Fig. 8a, b). Initially, we chemically linked fluorescent molecules and a peptide to CCMV protein cage platforms (Gillitzer et al. 2002). Fluorescent reporter molecules are useful for detection and quantification of the extent of derivatization. The degree of labeling of different sites on the CCMV platforms was highly dependent on reaction conditions (M.L. Flenniken et al., unpublished data; Gillitzer et al. 2002).

The small Hsp cage (*Mj*Hsp) of *M. jannaschii* was also labeled with fluorescent molecules, illustrating that many protein platforms within our library of cages perform similarly as chemical building blocks (Flenniken et al. 2003). Although

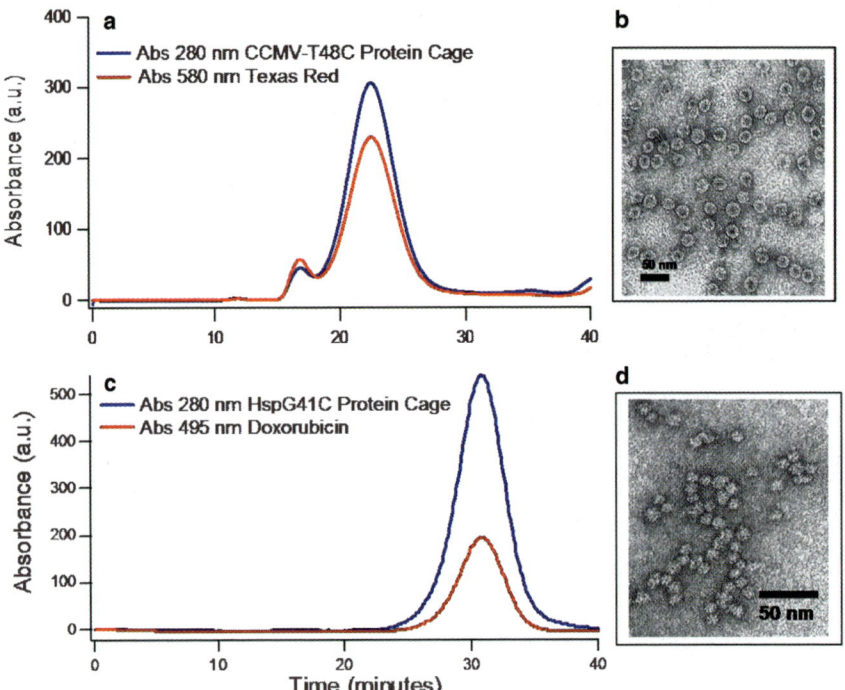

Fig. 8a–d Interiorly derivatized CCMV (T48C) and Hsp (G41C). **a** Size exclusion chromatography (SEC) elution profile of CCMV illustrating the co-elution of CCMV (T48C) protein cage (Abs 280 nm) and Texas Red (Abs 580 nm). **b** TEM of CCMV(T48C)-Texas red (stained with uranyl acetate). Scale bar, 50 nm. **c** SEC elution profile of Hsp (G41C) labeled with the (6-Maleimidocaproyl) hydrazone of doxorubicin (a chemotherapeutic agent), illustrating the co-elution of the Hsp protein cage (Abs 280) and doxorubicin (Abs 495). **d** TEM of HspG41C cages containing doxorubicin (stained with uranyl acetate). Scale bar, 50 nm. (Adapted from Flenniken et al. 2005)

similar in reactivity, the endogenous properties including structure, amino acid composition, size, pH stability, temperature stability, and solvent accessibility of each protein cage differs. These differences can be used to our advantage. The *Mj*Hsp cage as compared to CCMV is a much smaller (~12-nm exterior diameter), more solvent accessible protein cage assembled from 24 identical subunits. The *Mj*Hsp cage, like CCMV, is readily amenable to genetic modification. In order to chemically derivatize *Mj*Hsp in a spatially selective manner, we generated two mutants with either internally or externally exposed cysteine residues (G41C and S121C, respectively) and subsequently characterized their reactivities with activated fluorescent molecules (Flenniken et al. 2003). Protein architectures can be viewed as building blocks with a range of generic chemistry (e.g., thiol–maleimide coupling) that can be used to link reporter molecules and other compounds, including therapeutics. We demonstrated the selective attachment and release of a chemotherapeutic agent (doxorubicin) from the interior of a genetically modified *Mj*Hsp cage, HspG41C (Flenniken et al. 2005) (Fig. 8c, d). The advantage of this approach is that housing therapeutics within protein cages limits their bioavailability until their environmentally triggered release. We demonstrated pH-dependant release of doxorubicin from HspG41C under biologically relevant (lysosomal mimicking) conditions (Flenniken et al. 2005).

Another key component for the development of protein cage architectures as imaging and therapeutic agents is cell-specific targeting. In vivo application of the phage display library technique enabled the identification of peptides that bind specifically to the vasculature of particular organs as well as tumors (Arap et al. 1998a, 1998b, 2002; Pasqualini and Ruoslahti 1996a, 1996b; Pasqualini et al. 2000; Ruoslahti 2000). One of the most characterized of these targeting peptides is RGD-4C (CDCRGDCFC), which binds $\alpha_v\beta_3$ and $\alpha_v\beta_5$ integrins that are more prevalently expressed within tumor vasculature (Arap et al. 1998a; Brooks et al. 1994; Friedlander et al. 1995; Koivunen et al. 1995; Pasqualini et al. 1995). We incorporated RGD-4C and other targeting peptides on the exteriors of both *Mj*Hsp (HspG41C-RGD4C) and human H-chain ferritin (RGD4C-HFn) (Flenniken et al. 2006; Uchida et al. 2006). Fluorescein labeling of cell-specific targeted cages enables their visualization by epifluorescence microscopy. Using this approach, we demonstrated RGD-4C-mediated cell targeting of HspG41CRGD-4C cages in vitro (Fig. 9) (Flenniken et al. 2006). In addition to genetic incorporation, cell-specific targeting ligands, including antibodies and peptides, have also been chemically coupled to protein cage platforms (Chatterji et al. 2002, 2004b; Flenniken et al. 2006; Gillitzer et al. 2002; Medintz et al. 2005). For example, an anti-CD4 monoclonal antibody conjugated to fluorescently labeled HspG41C enabled targeting of CD4$^+$ lymphocytes within a population of splenocytes (Flenniken et al. 2006). The multivalent nature of protein cage architectures results in the presentation of multiple targeting ligands on their surfaces and may potentially aid in the interaction of these protein cages with many surfaces including receptors on a variety of cell types.

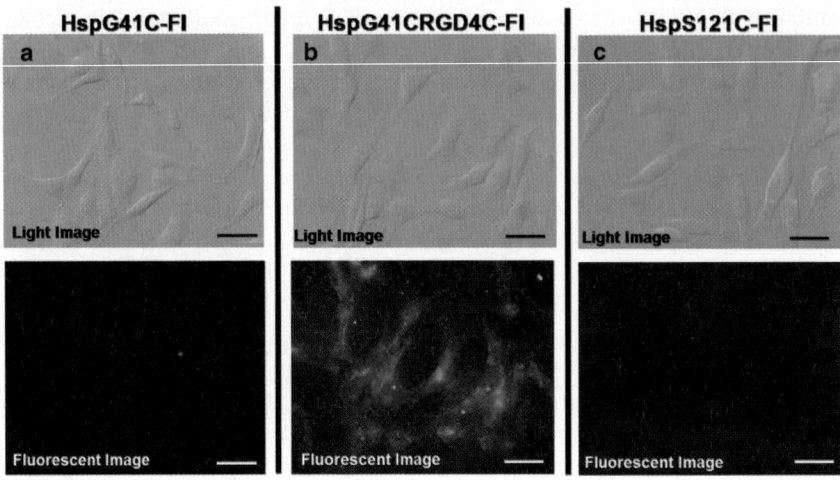

Fig. 9a–c Epifluorescence microscopy of C32 melanoma cells with Hsp cage-fluorescein conjugates. Cells were incubated with (**a**) nontargeted HspG41C-Fl cages with internally bound fluorescein, (**b**) tumor targeted HspG41CRGD4C-Fl cages and (**c**) nontargeted HspS121C-Fl cages with externally bound fluorescein. C32 melanoma cells grown on coverslips were incubated with Hsp cage-fluorescein conjugates and imaged by both light (*top*) and fluorescent microscopy (*bottom*). The fluorescein concentration for cage-cell incubations was 2.5 mM and all fluorescent images were taken at a standardized camera exposure time of 50 ms. Scale bar, 50 mm. (Flenniken et al. 2006)

Asymmetric Derivatization of Inherently Symmetric Protein Cage Architectures

High symmetry is an inherent property of protein cage architectures that are typically assembled from multiple copies of a single protein. While symmetric ligand presentation is often desired, we have also developed approaches for the asymmetric presentation of ligands on a protein cage while maintaining their structure (Gillitzer et al. 2006; Klem et al. 2003). In the first approach, one face of a CCMV capsid (genetic variant A163C with exteriorly exposed thiol groups) was reversibly bound through exposed thiols to an activated resin, followed by the passivation of the remaining thiols with iodoacetic acid and subsequent elution of the cages (Klem et al. 2003). The resulting CCMV capsids, with reactive thiol groups present on a single face, bound a solid gold substrate in an ordered fashion resulting in a 2D monolayer of virus particles (Klem et al. 2003). A second symmetry-breaking approach capitalized on the subunit assembly and disassembly process of protein cages to generate CCMV populations with asymmetric presentation of ligands (Gillitzer et al. 2006). Two populations of CCMV capsids differentially modified (with either biotin or digoxigenin) were disassembled into subunits and subsequently combined (in varying stoichiometries) and reassembled, resulting in dual functionalized capsids (Gillitzer et al. 2006). The introduction of dual functionality to a single protein cage platform is a step toward developing cages with multiple functionalities. Ultimately, the goal is to control the number and spatial arrangement of ligands on the cage, independent of the symmetry of the underlying architecture.

In Vivo Study of Protein Cage-Mediated Materials for Medical Applications

As described in the previous sections, protein cages have the potential as nanoplatforms in medical applications. However, it is critical to investigate biocompatibility and biodistribution of the protein cage particles in vivo before proceeding to preclinical and clinical studies. We have examined biodistribution of two different protein cages, CCMV and MjHsp and demonstrated both cages shows similar broad distribution in most tissues and organs and no obvious toxicity after a single injection (Kaiser et al. 2007). These results indicate that protein-cage-based nanoparticles are biocompatible and could be utilized as in vivo biomedical materials. On the basic of this knowledge, we have been developing iron-oxide-encapsulated protein cages for MRI contrast agents (M. Uchida et al., unpublished data).

Manchester and co-workers have studied the biocompatibility of another viral protein cage, CPMV, and revealed that it also has the potential to serve as a nontoxic platform in medical use (Rae et al. 2005; Singh et al. 2007). Furthermore, they have demonstrated that fluorescently labeled CPMV is internalized in vascular endothelial cells and can be used as an imaging probe to visualize the vasculature and blood flow in living mice (Lewis et al. 2006).

Introduction of Multiple Functionalities on a Single Protein Cage Architecture

Thus far, we have demonstrated functionalities including biomimetic mineralization, chemical derivatization with both fluorescent and therapeutic molecules, genetic and chemical introduction of cell-specific targeting ligands, genetic incorporation of metal coordinating peptides, and chemical linkage of gadolinium chelators on protein cage architectures. In addition, we have shown that protein cages with cell-specific targeting capabilities can simultaneously be chemically modified with fluorescent molecules, therefore creating a cage with both cell-targeting and imaging capabilities (Flenniken et al. 2006). The next phase of our efforts toward the development of protein cage architectures as nanomaterials for bioengineering and biomedicine is focused on the incorporation of multiple modalities on a single protein cage platform. Ultimately, the utility of these materials for medical applications must be evaluated in vivo and these studies are currently in progress.

Conclusion

The use of protein cage architectures of viral and nonviral origin as nanomaterials with applications in biomedicine and biotechnology provides a number of unique advantages. Their biological origin makes them both amenable to genetic modification and large-scale production. Genetic modification enables the site-specific introduction of chemical and/or structural functionality onto highly symmetric protein cage platforms. The presence of reactive functional groups also allows a chemical approach to the attachment and presentation of both organic and inorganic ligands. This structural and functional plasticity allows many protein cage systems within our library to be engineered and redesigned for specific applications in materials science, catalysis, and biomedicine. In addition, the ability to break the inherent symmetry of the cage-like architectures holds promise for very precise control over placement and presentation of functionalized ligands on these templates. The chemistry presented here is fundamentally biomimetic, where lessons learned from how biological systems deal with issues of spatial control and assembly have been applied to purely or partially synthetic systems. While there are certainly competing approaches, the use of protein cage architectures for nanomaterials synthesis is both an exciting and potentially fruitful arena.

References

Ackermann HW (2001) Frequency of morphological phage descriptions in the year 2000. Brief review. Arch Virol 146:843–857
Allen TM, Cullis PR (2004) Drug delivery systems: entering the mainstream. Science 303:1818–1822
Allen M, Willits D, Mosolf J, Young M, Douglas T (2002) Protein cage constrained synthesis of ferrimagnetic iron oxide nanoparticles. Adv Mater 14:1562–1565
Allen M, Willits D, Young M, Douglas T (2003) Constrained synthesis of cobalt oxide nanomaterials in the 12-subunit protein cage from *Listeria innocua*. Inorg Chem 42:6300–6305
Allen MA, Bulte JWM, Liepold L, Basu G, Zywicke HA, Frank JA, Young M, Douglas T (2005) Paramagnetic viral nanoparticles as potential high-relaxivity magnetic resonance contrast agents. Magn Reson Med 54:807–812
Anderson EA, Isaacman S, Peabody DS, Wang EY, Canary JW, Kirshenbaum K (2006) Viral nanoparticles donning a paramagnetic coat: conjugation of MRI contrast agents to the MS2 capsid. Nano Lett 6:1160–1164
Arap W, Pasqualini R, Ruoslahti E (1998a) Cancer treatment by targeted drug delivery to tumor vasculature in a mouse model. Science 279:377–380
Arap W, Pasqualini R, Ruoslahti E (1998b) Chemotherapy targeted to tumor vasculature. Curr Opin Oncol 10:560–565
Arap W, Haedicke W, Bernasconi M, Kain R, Rajotte D, Krajewski S, Ellerby HM, Bredesen DE, Pasqualini R, Ruoslahti E (2002) Targeting the prostate for destruction through a vascular address. Proc Natl Acad Sci U S A 99:1527–1531
Basu G, Allen M, Willits D, Young M, Douglas T (2003) Metal binding to cowpea chlorotic mottle virus using terbium(III) fluorescence. J Biol Inorg Chem 8:721–725
Belcher A, Flynn C, Whaley S, Mao CB, Gooch E (2001) Biomolecular recognition and control of nano magnetic and semiconductor materials. Abstr Papers Chemical Soc 222:53-POLY

Brooks PC, Montgomery AM, Rosenfeld M, Reisfeld RA, Hu T, Klier G, Cheresh DA (1994) Integrin alpha v beta 3 antagonists promote tumor regression by inducing apoptosis of angiogenic blood vessels. Cell 79:1157–1164

Brumfield S, Willits D, Tang L, Johnson JE, Douglas T, Young M (2004) Heterologous expression of modified Cowpea chlorotic mottle bromovirus coat protein results in the assembly of protein cages with altered architectures and function. J Gen Virol 85:1049–1053

Bulte JWM, Douglas T, Mann S, Frankel RB, Moskowitz BM, Brooks RA, Baumgarner CD, Vymazal J, Frank JA (1994a) Magnetoferritin: biomineralization as a novel approach in the design of iron oxide-based MR contrast agents. Inv Rad 29:S214–S216

Bulte JWM, Douglas T, Mann S, Frankel RB, Moskowitz BM, Brooks RA, Baumgarner CD, Vymazal J, Strub M-P, Frank JA (1994b) Magnetoferritin: characterization of a novel superparamagnetic MR contrast agent. J Magn Res Imaging 4:497–505

Bulte CJ, White O, Olsen GJ, Zhou L, Fleischmann RD, Sutton GG, Blake JA, FitzGerald LM, Clayton RA, Gocayne JD, Kerlavage AR, Dougherty BA, Tomb JF, Adams MD, Reich CI, Overbeek R, Kirkness EF, Weinstock KG, Merrick JM, Glodek A, Scott JL, Geoghagen NS, Venter JC (1996) Complete genome sequence of the methanogenic archaeon *Methanococcus jannaschii*. Science 273:1058–1073

Bulte JWM, Douglas T, Witwer B, Zhang SC, Strable E, Lewis BK, Zywicke H, Miller B, van Gelderen P, Moskowitz BM, Duncan ID, Frank JA (2001) Magnetodendrimers allow endosomal magnetic labeling and in vivo tracking of stem cells. Nat Biotechnol 19:1141–1147

Burritt JB, Bond CW, Doss KW, Jesaitis AJ (1996) Filamentous phage display of oligopeptide libraries. Anal Biotechnol. 238:1–13

Chasteen ND, Harrison PM (1999) Mineralization in ferritin: an efficient means of iron storage. J Struct Biol 126:182–194

Chatterji A, Burns LL, Taylor SS, Lomonossoff GP, Johnson JE, Lin T, Porta C (2002) Cowpea mosaic virus: from the presentation of antigenic peptides to the display of active biomaterials. Intervirology 45:362–370

Chatterji A, Ochoa WF, Paine M, Ratna BR, Johnson JE, Lin T (2004a) New addresses on an addressable virus nanoblock; uniquely reactive Lys residues on cowpea mosaic virus. Chem Biol 11:855–863

Chatterji A, Ochoa W, Shamieh L, Salakian SP, Wong SM, Clinton G, Ghosh P, Lin T, Johnson JE (2004b) Chemical conjugation of heterologous proteins on the surface of Cowpea mosaic virus. Bioconjug Chem 15:807–813

Chatterji A, Ochoa WF, Ueno T, Lin T, Johnson JE (2005) A virus-based nanoblock with tunable electrostatic properties. Nano Lett 5:597–602

Datta A, Hooker JM, Botta M, Francis MB, Aime S, Raymond KN (2008) High relaxivity gadolinium hydroxypyridonate-viral capsid conjugates: nanosized MRI contrast agents. 1. J Am Chem Soc 130:2546–2552

Douglas T (1996) Biomimetic synthesis of nanoscale particles in organized protein cages. In: Mann S (ed) Biomimetic approaches in materials science. VCH, New York, pp 91–115

Douglas T, Mann S (1995) Biomolecules in the synthesis of inorganic solids. In: Meyers RA (ed) Molecular biology and biotechnology. VCH, New York, pp 466–469

Douglas T, Ripoll D (1998) Electrostatic gradients in the iron storage protein ferritin. Protein Sci 7:1083–1091

Douglas T, Young M (1998) Host-guest encapsulation of materials by assembled virus protein cages. Nature (London) 393:152–155

Douglas T, Young M (1999) Virus particles as templates for materials synthesis. Adv Mater 11:679–681

Douglas T, Stark VT (2000) Nanophase cobalt oxyhydroxide mineral synthesized within the protein cage of ferritin. Inorg Chem 39:1828–1830

Douglas T, Allen M, Young M (2002a) Self-assembling protein cage systems and applications in nanotechnology. In: Fahnstock SR, Steinbuchel A (eds) Polyamides and complex proteinaceous materials I, Vol. 7. Wiley-VCH, Weinheim, p 517

Douglas T, Strable E, Willits D, Aitouchen A, Libera M, Young M (2002b) Protein engineering of a viral cage for constrained nano-materials synthesis. Adv Mater 14:415–418

Douglas T, Allen M, Klem M, Gilmore K, Idzerda Y, Young M (2004) Engineered protein cages for nanomaterials. Abstr Papers Am Chem Soc 227:U519–U519

Endo M, Fujitsuka M, Majima T (2007) Porphyrin light-harvesting arrays constructed in the recombinant tobacco mosaic virus scaffold. Chem Eur J 13:8660–8666

Ensign D, Young MJ, Douglas T (2004) Photocatalytic synthesis of copper colloids from Cu(II) by the ferrihydrite core of ferritin. Inorg Chem 43:3441–3446

Flenniken ML, Willits DA, Brumfield S, Young MJ, Douglas T (2003) The small heat shock protein cage from *Methanococcus jannaschii* is a versatile nanoscale platform for genetic and chemical modification. Nano Lett 3:1573–1576

Flenniken ML, Allen M, Young M, Douglas T (2004) Viruses as host assemblies. In: Encyclopedia of supramolecular chemistry. Steed AJW (ed) NMarcel Dekker, ew York City, pp 1563–1568

Flenniken ML, Liepold LO, Crowley BE, Willits DA, Young MJ, Douglas T (2005) Selective attachment and release of a chemotherapeutic agent from the interior of a protein cage architecture. Chem Commun (Camb) 447–449

Flenniken ML, Willits DA, Harmsen AL, Liepold LO, Harmsen AG, Young MJ, Douglas T (2006) Melanoma and lymphocyte cell-specific targeting incorporated into a heat shock protein cage architecture. Chem Biol 13:161–170

Friedlander M, Brooks PC, Shaffer RW, Kincaid CM, Varner JA, Cheresh DA (1995) Definition of two angiogenic pathways by distinct alpha v integrins. Science 270:1500–1502

Gillitzer E, Willits D, Young M, Douglas T (2002) Chemical modification of a viral cage for multivalent presentation. Chem Commun (Camb) 2390–2391

Gillitzer E, Succi P, Young M, Douglas T (2006) Controlled ligand display on a symmetrical protein-cage architecture through mixed assembly. Small 2:962–966

Harrison PM, Arosio P (1996) The ferritins: molecular properties, iron storage function and cellular regulation. Biochim Biophys Acta 1275:161–203

Harrison PM, Banyard SH, Hoare RJ, Russell SM, Treffry A (1976) The structure and function of ferritin. Ciba Found Symp 19–40

Hashizume H, Baluk P, Morikawa S, McLean JW, Thurston G, Roberge S, Jain RK, McDonald DM (2000) Openings between defective endothelial cells explain tumor vessel leakiness. Am J Pathol 156:1363–1380

Hikono T, Matsumura T, Miura A, Uraoka Y, Fuyuki T, Takeguchi M, Yoshii S, Yamashita I (2006) Electron confinement in a metal nanodot monolayer embedded in silicon dioxide produced using ferritin protein. Appl Phys Lett 88:023108

Hooker JM, Kovacs EW, Francis MB (2004) Interior surface modification of bacteriophage MS2. J Am Chem Soc 126:3718–3719

Hooker JM, Datta A, Botta M, Raymond KN, Francis MB (2007) Magnetic resonance contrast agents from viral capsid shells: a comparison of exterior and interior cargo strategies. Nano Lett 7:2207–2210

Hosein HA, Strongin DR, Allen M, Douglas T (2004) Iron and cobalt oxide and metallic nanoparticles prepared from ferritin. Langmuir 20:10283–10287

Ichikawa K, Uraoka Y, Punchaipetch P, Yano H, Hatayama T, Fuyuki T, Yamashita I (2007) Low-temperature polycrystalline silicon thin film transistor flash memory with ferritin. Jpn J Appl Phys 46:L804–L806

Iwahori K, Yoshizawa K, Muraoka M, Yamashita I (2005) Fabrication of ZnSe nanoparticles in the apoferritin cavity by designing a slow chemical reaction system. Inorg Chem 44:6393–6400

Johnson JE, Speir JA (1997) Quasi-equivalent viruses: a paradigm for protein assemblies. J Mol Biol 269:665–675

Kaiser CR, Flenniken ML, Gillitzer E, Harmsen AL, Harmsen AG, Jutila MA, Douglas T, Young MJ (2007) Biodistribution studies of protein cage nanoparticles demonstrate broad tissue distribution and rapid clearance in vivo. Int J Nanomed 2:715–733

Kawashita M, Tanaka M, Kokubo T, Inoue Y, Yao T, Hamada S, Shinjo T (2005) Preparation of ferrimagnetic magnetite microspheres for in situ hyperthermic treatment of cancer. Biomater 26:2231–2238

Kim I, Hosein HA, Strongin DR, Douglas T (2002) Photochemical reactivity of ferritin for Cr(VI) reduction. Chem Mater 14:4874–4879

Kim KK, Kim R, Kim SH (1998a) Crystal structure of a small heat-shock protein. Nature 394:595–599

Kim KK, Yokota H, Santoso S, Lerner D, Kim R, Kim SH (1998b) Purification, crystallization, and preliminary X-ray crystallographic data analysis of small heat shock protein homolog from *Methanococcus jannaschii*, a hyperthermophile. J Struct Biol 121:76–80

Klem MT, Willits D, Young M, Douglas T (2003) 2-D array formation of genetically engineered viral cages on Au surfaces and imaging by atomic force microscopy. J Am Chem Soc 125:10806–10807

Klem M, Willits D, Solis DJ, Belcher A, Young M, Douglas T (2005a) Bio-inspired synthesis of protein-encapsulated CoPt nanoparticles. Adv Funct Mater 15:1489–1494

Klem M, Young M, Douglas T (2005b) Biomimetic magnetic nanoparticles. Mater Today 8:28–37

Klem MT, Mosolf J, Young M, Douglas T (2008) Photochemical mineralization of europium titanium, iron oxyhydroxide nanoparticles in the ferritin protein cage. Inorg Chem 47:2237–2239

Koivunen E, Wang B, Ruoslahti E (1995) Phage libraries displaying cyclic peptides with different ring sizes: ligand specificities of the RGD-directed integrins. Biotechnology (N Y) 13:265–270

Kramer RM, Li C, Carter DC, Stone MO, Naik RR (2004) Engineered protein cages for nanomaterial synthesis. J Am Chem Soc 126:13282–13286

Le Clainche L, Plancque G, Amekraz B, Moulin C, Pradines-Lecomte C, Peltier G, Vita C (2003) Engineering new metal specificity in EF-hand peptides. J Biol Inorg Chem 8:334–340

Lewis JD, Destito G, Zijlstra A, Gonzalez MJ, Quigley JP, Manchester M, Stuhlmann H (2006) Viral nanoparticles as tools for intravital vascular imaging. Nat Med 12:354–360

Liepold LO, Willits D, Oltrogge L, Allen M, Young M, Douglas T (2007) Viral capsids as MRI contrast agents. Magn Reson Med 58:871–879

Luderer AA, Borrelli NF, Panzarino JN, Mansfield GR, Hess DM, Brown JL, Barnett EH, Hahn EW (1983) Glass-ceramic-mediated, magnetic-field-induced localized hyperthermia: response of a murine mammary carcinoma. Radiat Res 94:190–198

Maeda H, Wu J, Sawa T, Matsumura Y, Hori K (2000) Tumor vascular permeability and the EPR effect in macromolecular therapeutics: a review. J Control Release 65:271–284

Maeder DL, Weiss RB, Dunn DM, Cherry JL, Gonzalez JM, DiRuggiero J, Robb FT (1999) Divergence of the hyperthermophilic archaea *Pyrococcus furiosus* and *P. horikoshii* inferred from complete genomic sequences. Genetics 152:1299–1305

Mao C, Flynn CE, Hayhurst A, Sweeney R, Qi J, Georgiou G, Iverson B, Belcher AM (2003) Viral assembly of oriented quantum dot nanowires. Proc Natl Acad Sci U S 100:6946–6951

Mao C, Solis DJ, Reiss BD, Kottmann ST, Sweeney RY, Hayhurst A, Georgiou G, Iverson B, Belcher AM (2004) Virus-based toolkit for the directed synthesis of magnetic and semiconducting nanowires. Science 303:213–217

Matsui T, Matsukawa N, Iwahori K, Sano KI, Shiba K, Yamashita I (2007) Direct production of a two-dimensional ordered array of ferritin-nanoparticles on a silicon substrate. Jpn J Appl Phys 46:L713–L715

Medintz IL, Sapsford KE, Konnert JH, Chatterji A, Lin T, Johnson JE, Mattoussi H (2005) Decoration of discretely immobilized cowpea mosaic virus with luminescent quantum dots. Langmuir 21:5501–5510

Miller RA, Presley AD, Francis MB (2007) Self-assembling light-harvesting systems from synthetically modified tobacco mosaic virus coat proteins. J Am Chem Soc 129:3104–3109

Miura A, Hikono T, Matsumura T, Yano H, Hatayama T, Uraoka Y, Fuyuki T, Yoshii S, Yamashita I (2006) Floating nanodot gate memory devices based on biomineralized inorganic nanodot array as a storage node. Jpn J Appl Phys 45:L1–L3

Narberhaus F (2002) Alpha-crystallin-type heat shock proteins: socializing minichaperones in the context of a multichaperone network. Microbiol Mol Biol Rev 66:64–93

Nkere UU, Walter NM, Nikandrov VV, Gratzel CK, Moser JE, Gratzel MJ (1997) Light induced redox reactions involving mammalian ferritin as a photocatalyst. Photochem Photobiol B 41:83–89

Parker MJ, Allen MA, Ramsay B, Klem MT, Young M, Douglas T (2008) Expanding the temperature range of biomimetic synthesis using a ferritin from the hyperthermophile *Pyrococcus furiosus*. Chem Mater 20:1541–1547

Pasqualini R, Ruoslahti E (1996a) Organ targeting in vivo using phage display peptide libraries. Nature 380:364–366

Pasqualini R, Ruoslahti E (1996b) Tissue targeting with phage peptide libraries. Mol Psychiatry 1:423

Pasqualini R, Koivunen E, Ruoslahti E (1995) A peptide isolated from phage display libraries is a structural and functional mimic of an RGD-binding site on integrins. J Cell Biol 130:1189–1196

Pasqualini R, Koivunen E, Kain R, Lahdenranta J, Sakamoto M, Stryhn A, Ashmun RA, Shapiro LH, Arap W, Ruoslahti E (2000) Aminopeptidase N is a receptor for tumor-homing peptides and a target for inhibiting angiogenesis. Cancer Res 60:722–727

Prasuhn DE, Yeh RM, Obenaus A, Manchester M, Finn MG (2007) Viral MRI contrast agents: coordination of Gd by native virions and attachment of Gd complexes by azide-alkyne cycloaddition. Chem Comm 28:1269–1271

Rae CS, Khor IW, Wang Q, Destito G, Gonzalez MJ, Singh P, Thomas DM, Estrada MN, Powell E, Finn MG, Manchester M (2005) Systemic trafficking of plant virus nanoparticles in mice via the oral route. Virology 343:224–235

Raja KS, Wang Q, Finn MG (2003a) Icosahedral virus particles as polyvalent carbohydrate display platforms. Chembiochem 4:1348–1351

Raja KS, Wang Q, Gonzalez MJ, Manchester M, Johnson JE, Finn MG (2003b) Hybrid virus-polymer materials. 1. Synthesis and properties of PEG-decorated cowpea mosaic virus. Biomacromolecules 4:472–476

Ramsay B, Wiedenheft B, Allen M, Gauss GH, Lawrence CM, Young M, Douglas T (2006) Dps-like protein from the hyperthermophilic archaeon *Pyrococcus furiosus*. J Inorg Biochem 100:1061–1068

Reddy VS, Nataraja, P, Okerberg B, Li K, Damodaran KV, et al (2001) Virus Partilce Explorer (VIPER), a website for virus capsid structures and their computational analyses. J Virol 75:11943–11947

Rice G, Stedman K, Snyder J, Wiedenheft B, Willits D, Brumfield S, McDermott T, Young MJ (2001) Viruses from extreme thermal environments. Proc Natl Acad Sci U S A 98:13341–13345

Rice G, Tang L, Stedman K, Roberto F, Sphuler J, Gillitzer E, Johnson JE, Douglas T, Young M (2004) The structure of a thermophilic archaeal virus shows a double-stranded DNA viral capsid type that spans all three domains of life. Proc Natl Acad Sci U S A 101:7716–7720

Ruoslahti E (2000) Targeting tumor vasculature with homing peptides from phage display. Semin Cancer Biol 10:435–442

Schlick TL, Ding Z, Kovacs EW, Francis MB (2005) Dual-surface modification of the tobacco mosaic virus. J Am Chem Soc 127:3718–3723

Schneemann A, Young MJ (2003) Viral assembly using heterologous expression systems and cell extracts. Adv Protein Chem 64:1–36

Scott JK, Smith GP (1990) Searching for peptide ligands with an epitope library. Science 249:386–390

Seeman NC, Belcher AM (2002) Emulating biology: Building nanostructures from the bottom up. Proc Natl Acad Sci U S A 99:6451–6455

Singh P, Prasuhn D, Yeh RM, Destito G, Rae CS, Osborn K, Finn MG, Manchester M (2007) Bio-distribution, toxicity and pathology of cowpea mosaic virus nanoparticles in vivo. J Control Release 120:41–50

Smith GP, Petrenko VA (1997) Phage display. Chem Rev 97:391–410

Snyder JC, Stedman K, Rice G, Wiedenheft B, Spuhler J, Young MJ (2003) Viruses of hyperthermophilic Archaea. Res Microbiol 154:474–482

Speir JA, Munshi S, Wang G, Baker TS, Johnson JE (1995) Structures of the native and swollen forms of cowpea chlorotic mottle virus determined by X-ray crystallography and cryo-electron microscopy. Structure 3:63–78

Stefanini S, Cavallo S, Montagnini B, Chiancone E (1999) Incorporation of iron by the unusual dodecameric ferritin from *Listeria innocua*. Biochem J 338:71–75

Strable E, Johnson JE, Finn MG (2004) Natural nanochemical building blocks: icosahedral virus particles organized by attached oligonucleotides. Nano Lett 4:1385–1389

Su M, Cavallo S, Stefanini S, Chiancone E, Chasteen ND (2005) The so-called *Listeria innocua* ferritin is a Dps protein. Iron incorporation, detoxification, DNA protection properties. Biochemistry 44:5572–5578

Suci PA, Berglund DL, Liepold L, Brumfield S, Pitts B, Davison W, Oltrogge L, Hoyt KO, Codd S, Stewart PS, Young M, Douglas T (2007) High-density targeting of a viral multifunctional nanoplatform to a pathogenic, biofilm-forming bacterium. Chem Biol 14:387–398

Uchida M, Flenniken ML, Allen M, Willits DA, Crowley BE, Brumfield S, Willis AF, Jackiw L, Jutila M, Young MJ, Douglas T (2006) Targeting of cancer cells with ferrimagnetic ferritin cage nanoparticles. J Am Chem Soc 128:16626–16633

Varpness Z, Peters JW, Young M, Douglas T (2005) Biomimetic synthesis of a H2 catalyst using a protein cage architecture. Nano Lett 5:2306–2309

Wang Q, Kaltgrad E, Lin T, Johnson JE, Finn MG (2002a) Natural supramolecular building blocks. Wild-type cowpea mosaic virus. Chem Biol 9:805–811

Wang Q, Lin T, Johnson JE, Finn MG (2002b) Natural supramolecular building blocks. Cysteine-added mutants of cowpea mosaic virus. Chem Biol 9:813–819

Wang Q, Lin TW, Tang L, Johnson JE, Finn MG (2002c) Icosahedral virus particles as addressable nanoscale building blocks. Angew Chem Int Ed. 41:459–462

Whaley SR, English DS, Hu EL, Barbara PF, Belcher AM (2000) Selection of peptides with semiconductor binding specificity for directed nanocrystal assembly. Nature 405:665–668

Wiedenheft B, Mosolf J, Willits D, Yeager M, Dryden KA, Young M, Douglas T (2005) From the cover: an archaeal antioxidant: characterization of a Dps-like protein from *Sulfolobus solfataricus*. Proc Natl Acad Sci U S A 102:10551–10556

Wikipedia (2005) Wikipedia Free Encyclopedia, Wikipedia

Yamada K, Yoshii S, Kumagai S, Miura A, Uraoka Y, Fuyuki T, Yamashita I (2007) Effects of dot density and dot size on charge injection characteristics in nanodot array produced by protein supramolecules. Jpn J Appl Phys 46:7549–7553

Yang XK, Chiancone E, Stefanini S, Ilari A, Chasteen ND (2000) Iron oxidation and hydrolysis reactions of a novel ferritin from *Listeria innocua*. Biochem J 349:783–786

Zhao X, Fox JM, Olson NH, Baker TS, Young MJ (1995) In vitro assembly of cowpea chlorotic mottle virus from coat protein expressed in *Escherichia coli* and in vitro-transcribed viral cDNA. Virology 207:486–494

Biomedical Nanotechnology Using Virus-Based Nanoparticles

G. Destito, A. Schneemann, M. Manchester (✉)

Contents

Why Do We Need Biomedical Nanotechnology? 96
   Types of Nanoparticles in Use 96
   Viruses as Natural Nanomaterials 97
Virus Particles We Use for Biomedical Applications 98
   Cowpea Mosaic Virus 98
   Flock House Virus 102
Applications of CPMV and FHV Particles to Therapeutics and Tumor Targeting 109
   Tumor Targeting 109
   Vaccine Applications for CPMV and FHV 113
   Virus-Based Therapeutics: Antivirals and Antitoxins 115
Advantages and Disadvantages Comparing CPMV and FHV Type Systems 116
The Future of Viral Nanoparticles for in Vivo Therapeutic and Diagnostic Purposes ... 116
   Evaluating the Biodistribution of VNPs in Vivo 116
   Immunologic Properties of CPMV in Vivo 117
   Toxicity Studies of CPMV in Vivo 118
Conclusions 118

Abstract A great challenge in biomedicine is the ability to target therapeutics to specific locations in the body in order to increase therapeutic benefit and minimize adverse effects. Virus-based nanotechnology takes advantage of the natural circulatory and targeting properties of viruses, in order to design therapeutics and vaccines that specifically target tissues of interest in vivo. Cowpea mosaic virus (CPMV) and flock house virus (FHV) nanoparticle-based strategies hold great promise for the design of targeted therapeutics, as well as for structure-based vaccine approaches.

Abbreviations APC Antigen-presenting cell, CCMV Cowpea chlorotic mottle virus, CEA Carcinoembryonic antigen, CPMV Cowpea mosaic virus, CTL Cytotoxic T-lymphocyte, cDNA Complementary deoxyribonucleic acid, DC Dendritic cell,

M. Manchester
Department of Cell Biology, Center for Integrative Molecular Biosciences, The Scripps Research Institute, CB262, 10550 N. Torrey Pines Road, La Jolla, CA 92037, USA
e-mail: marim@scripps.edu

DMSO Dimethyl sulfoxide, ELISA Enzyme-linked immunosorbent assay, FA Folic acid, FHV Flock house virus, MHC Major histocompatibility complex, MRI Magnetic resonance imaging, NHS N-hydroxysuccinimide, NPY Neuropeptide Y, QD Quantum dots, RNA Ribonucleic acid, SWCNT Single-walled carbon nanotubes, USPIO Ultra-small paramagnetic iron oxide, UV Ultraviolet, VNP Viral nanoparticles, VLP Virus-like particle

Why Do We Need Biomedical Nanotechnology?

The past quarter-century of outstanding progress in basic biomedical research is only beginning to be translated into comparable advances in the clinic. Arguably the tools for best applying advances based on molecular principles exist at the nanoscale. Recently, nanotechnology research has begun making tremendous progress in the biomedical field, and researchers are now focusing their attention on analyzing the events at the molecular and cellular level.

One example where rapid and significant progress is urgently needed is in the early detection and treatment of tumors. The main drawback in using conventional chemotherapeutic drugs is related to the fact that only a relatively small amount of the administered substances reaches the cancer cells. The range and degree of the devastating and invasive side effects of chemotherapy are tragically well known. One of the new challenges in cancer medicine is the ability to target therapeutics to specific locations in the body in order to avoid adverse side effects that accompany the large systemic doses required of a nontargeted chemotherapeutic agent. There are also many effective chemotherapeutics whose clinical use has not been realized because of their systemic toxicity. An ideal therapeutic system would be selectively targeted against cell clusters that are in the early stages of the transformation toward the malignant phenotype (reviewed in Ferrari 2005).

If properly integrated with established cancer research, nanotechnology can be a powerful tool in the hands of researchers facing many challenges such as the identification of suitable early markers of neoplastic disease, along with the development of biomarker-targeted delivery of multiple therapeutic agents. An obvious advantage of nanotechnology as it relates to biological systems is the ability to control the size of the resulting particles and devices. The design of a multifunctional nanoparticle capable of targeting a specific tissue or cell type, delivering a contrast agent to allow for noninvasive imaging and a therapeutic payload to the target will revolutionize the traditional methods of cancer detection and treatment. Biologists, physicists and chemists, armed with new and effective tools provided by nanotechnology, have now joined forces to face this problem.

Types of Nanoparticles in Use

Nanotechnology concerns the study of devices that are themselves or have essential components in the 1- to 1,000-nm dimensional range. Because of their small size, nanoscale devices can readily interact with biomolecules on both the surface of cells and inside cells. By gaining access to so many areas of the body, they have the potential to detect disease and deliver treatment in ways unimagined before now. Liposomes, 50–70 nm in diameter, are the archetypal, simplest form of a nanoparticle. Liposome-encapsulated formulations of the chemotherapeutic doxorubicin were approved 10 years ago for the treatment of Kaposi's sarcoma, and are now commonly used against other types of cancer. Several types of nanoparticles have already demonstrated use for the enhancement of MRI contrast in imaging tumors. Ultra-small superparamagnetic iron oxide (USPIO) nanoparticles, coated with dextran, have been recently used to image lymph nodes containing micrometastases in patients with prostate cancer (Harisinghani et al. 2003). Other studies have used gadolinium or iron-oxide-based nanoparticles that combine magnetic resonance with biological targeting (Levy 2002).

Targeting of cancer cells and drug delivery has also been investigated using highly branched macromolecules called dendrimers. Dendrimers are synthesized through a repetitive reaction sequence that guarantees complete shells for each generation, leading to polymers that are monodisperse (Morawski et al. 2004). The synthetic procedures developed for dendrimer preparation allow for nearly complete control over the critical molecular design parameters, such as size, shape, surface/interior chemistry, and topology (Quintana 2002; Morawski et al. 2004). The problem is that these molecules possess an enormous number of energetically permissible conformations, and in solution there is rapid interchange between them. Nevertheless, dendrimers are presently at the forefront of nanoparticles used for many applications, including tumor targeting (Quintana 2002). Other examples of nanoparticle formulations include gold nanoshells, coated with cancer cell-specific ligands, which can act as molecularly targeted contrast agents for optical imaging; quantum dots, which have been used to label and track multiple cell types as they interact with one another in vivo; and hydrogels, very large molecules with complex three-dimensional structures capable of storing either small molecule drugs or much larger peptide and protein therapeutics. Several of these strategies have met with difficulties in vivo: for example quantum dots, while highly effective for imaging, have problems with toxicity of the heavy metals they are fabricated from. Similarly for hydrogels, bioelimination is a challenge. Once all of the drug has been delivered, hydrogels remain in the body unless surgically removed, as they cannot be broken down and eliminated. Thus there remains a great need for novel formulations of nanoparticles that are biocompatible and nontoxic.

Viruses as Natural Nanomaterials

Viruses are an excellent example of nanomaterials because of their regular geometries, well-characterized surface properties, and uniformity of size, making them ideal platforms for nanoscale fabrication (Brumfield et al. 2004; Klem et al. 2003; Singh et al. 2005). Many nanotechnology applications, including biomaterials, vaccines, chemical tools, and molecular electronic materials, have been approached using viruses (reviewed in Singh et al. 2005). Examples of viruses or virus-sized multimeric protein assemblies (termed protein cages) being developed for these purposes include cowpea chlorotic mottle virus (CCMV), cowpea mosaic virus (CPMV), flock house virus (FHV), vault nanocapsules, ferritin, hepatitis B cores, and bacteriophages such as M13, MS2 and Qβ (Chatterji et al. 2002, 2004a; Douglas and Young 1998, 1999; Flenniken et al. 2006; Hooker et al. 2004; Kickhoefer et al. 2005; Lee et al. 2002; Manayani et al. 2007; Mao et al. 2003, 2004; Prasuhn et al. 2008). Because of the extensive molecular and structural studies done on many of these viruses and virus-shaped particles, researchers can identify or modify amino acids in the viral capsid for bioconjugation, allowing for chemical modification of those amino acids both on the outer and inner capsid. Also, the capability for genetic modification and knowledge of the capsid structure allows sophisticated rational design of the ligand orientation and stoichiometry in order to maximize target recognition. A significant advantage of virus-based nanoplatforms as opposed to other backbones is that there are multiple sites on the virus capsid that can be used for the introduction of foreign peptides or ligands, so that a large number of ligands can be displayed in a precise spatial distribution. Achieving this level of control has not been possible with inorganic or lipid materials. The creation of a polyvalent scaffold that enhances the targeting properties of the virus by several orders of magnitude offers the possibility of cell targeting using much lower concentrations of targeting molecules than would be possible for the monomeric form (Sen Gupta et al. 2005b).

Virus Particles We Use for Biomedical Applications

Cowpea Mosaic Virus

Cowpea mosaic virus (CPMV) is a 31-nm, icosahedral plant virus that grows in the common cowpea plant (*Vigna unguiculata*) (Lomonossoff and Johnson 1991). CPMV has a bipartite positive-sense, single-stranded RNA genome with each RNA molecule (designated RNA1 and RNA2) encapsidated in a separate particle. Both RNA molecules are required for infection of plants, and infectious cDNA clones are available in the laboratory (Lomonossoff et al. 1993). The replication machinery and viral proteases are encoded by RNA1, while RNA2 encodes the movement protein (for transport of virions from cell to cell in the plant) and viral capsid proteins. Mutagenesis of capsid proteins encoded by RNA2 enables multivalent display of single or multiple amino acid substitutions and insertions on the particle surface.

CPMV capsids are composed of 60 copies each of a large (L; 42-kD) and small (S; 24-kD) capsid protein to form a 31-nm-diameter pseudo T=3 icosahedral particle. Figure 1 illustrates the organization of the CPMV capsid as a space-filling drawing of a wild type CPMV virion with the L and S subunits indicated. Structural comparisons of the capsids between CPMV and other related viruses showed significant

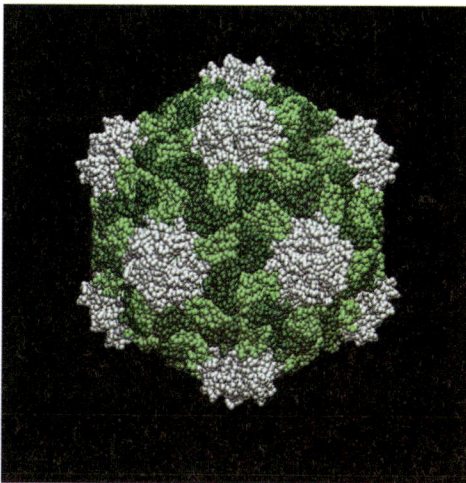

Fig. 1 A space-filling model of the CPMV capsid showing the L subunit in *dark green* and *light green spheres*, and the S subunit shown in *grey spheres*

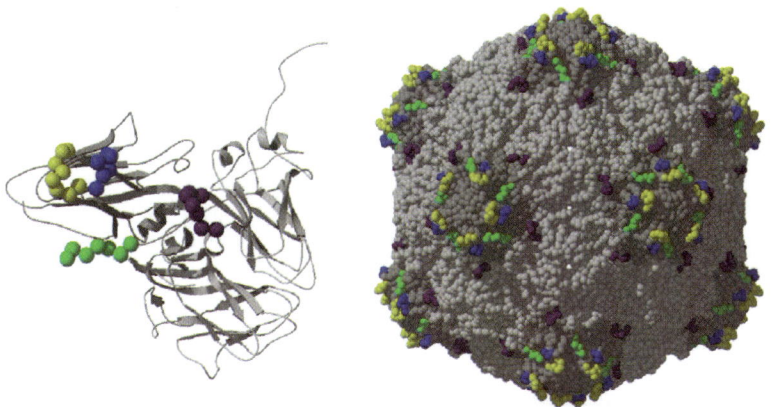

Fig. 2 *Left panel* Asymmetric unit (one L and one S subunit) of the CPMV capsid showing locations of insertion sites on the capsid proteins. On the S subunit, the βB-βC loop is in highlighted in *yellow*, the CÎ-C″ loop in *blue*, the S-subunit C-terminus in *green*, and on the L subunit the βE-βF loop in *purple*. *Right panel* Space-filling model of the whole CPMV capsid showing the locations of the same insertion sites on the capsid surface. There are 60 copies of each of these sites per particle

conformational variation in certain regions that are exposed on the surface of the particle (Lin 1999). Detailed examination of the three-dimensional structure of CPMV solved in Dr. Jack Johnson's laboratory revealed that there are several exposed beta-loops on the surface, termed βB-βC (small subunit), βE-αF, and C′-C″ (large subunit), or the C-terminus of the small subunit (Lin 1999) (Fig. 2). Subsequent mutagenesis has confirmed that these locations can accommodate additional peptide sequences or modifications (Porta 1996). The particle interior may also be modified genetically to introduce peptides and relatively large protein domains (up to 100 amino acids (A. Chatterji and J. Johnson, unpublished data).

CPMV production from experimentally infected plants does not require sterile culture techniques or costly reagents such as culture medium or serum. In addition, CPMV is nonpathogenic for humans, and the products derived from plant virus culture are not contaminated with animal cells or viruses (Brennan et al. 2001). CPMV spreads first within the inoculated leaves of an experimentally infected cowpea seedling and then moves systemically throughout the plant, with virus harvest occurring approximately 10 days later. CPMV grows to very high quantities in infected plants; approximately 1 g/kg of purified leaves is a typical yield (Johnson et al. 1997). Infected plants can be grown indoors using artificial light, although yields are improved with greenhouse conditions. CPMV purification is easy to perform and can be completed in a short time (Johnson et al. 1997). CPMV particles are very stable and can withstand extremes of temperature and pH while retaining activity (Lomonossoff and Johnson 1991). Inactivating the infectivity of CPMV particles while retaining their other materials properties can be accomplished using several inactivation strategies, including UV irradiation (Phelps et al. 2007), (C. Rae and M. Manchester, unpublished data).

Chemical Biology of CPMV: Direct Attachment of Functional Groups to the Particle Surface

Because of possible constraints on the replication of CPMV-containing surface capsid modifications, it can be difficult to recover viruses displaying large (>30 amino acids) or highly hydrophobic inserts at high yields. An alternative method for presenting proteins or peptides on CPMV is via direct chemical attachment to both wild- ype and mutated virus-based nanoparticles (VNPs) by straightforward chemistry that was developed by the Finn and Johnson laboratories at TSRI using fluorescent dyes (Wang et al. 2002a, 2002b). CPMV is stable to a variety of reaction conditions, including acid and basic pH, and organic solvents such as DMSO. Wild type CPMV has five surface lysine residues per asymmetric unit, which react with fluorescent dye-labeled N-hydroxysuccinimide (NHS) esters, as shown in Fig. 3. At neutral pH, wild type CPMV reacts with NHS ester at room temperature, up to a ratio of approximately 70–140 dye molecules per virion. The reactivity of each of the individual surface lysines has recently been determined where it appears that lysines 38 and 199 are the main targets (Chatterji et al. 2004a). Mutants are also available with selected lysines removed for even more precise control of ligand attachment and spatial arrangements (Chatterji et al. 2004a).

Wild type CPMV lacks addressable free thiol residues on the capsid surface, but there are approximately 30 accessible thiols (cysteine residues) on the capsid interior that can be targeted by maleimide-linked dyes. On the surface additional thiols have also been inserted via mutagenesis in the βB-βC or βE-αF loops. Attachment to surface thiols has been accomplished via maleimide-containing linkers under

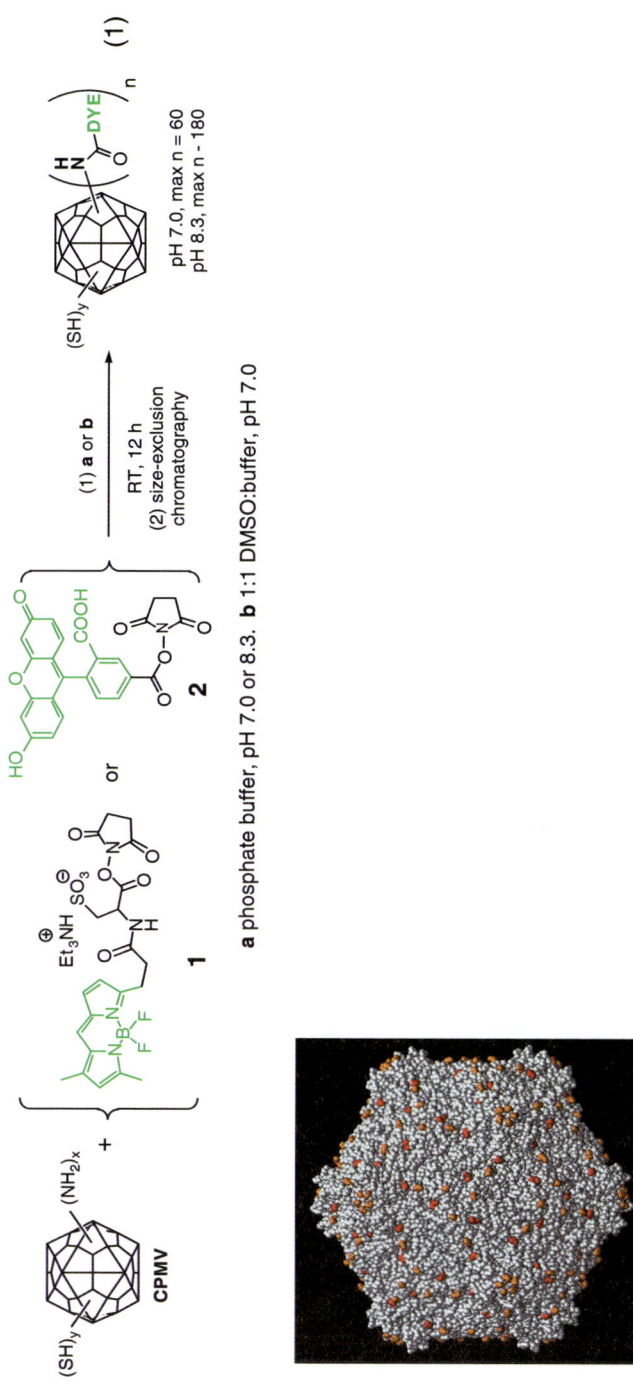

Fig. 3 *Top* Covalent attachment of fluorescein-NHS ester dye to wild type CPMV. *Bottom* Locations of lysines on the CPMV capsid at residues S38 (*red*), S82 (*orange*), L19 (*orange*), L34 (*orange*), L199 (*red*)

straightforward reaction conditions (Wang et al. 2002b). Through these two strategies, a variety of proteins have been conjugated (Chatterji et al. 2004b). An alternative strategy has recently been developed by the Finn laboratory using azide-alkyne "click" chemistry to efficiently attach ligands onto the CPMV surface (Destito 2007; Sen Gupta et al. 2005a; Wang et al. 2003). Together these chemical methods have expanded the reactivity of CPMV and provided us with a wider variety of tools for displaying ligands on the particle surface with precise control over their spatial distribution and orientation. This feature is unique to virus and protein-cage nanomaterials and is not found with other types of nanoparticles.

Flock House Virus

Flock house virus (FHV) is a member of the family *Nodaviridae*, which are small nonenveloped, T=3 icosahedral viruses with a bipartite, positive-sense RNA genome. The *Nodaviridae* are divided into two genera: alphanodaviruses, which infect insects and betanodaviruses, which infect fish. FHV is an alphanodavirus that was originally isolated from a grass grub in New Zealand. The FHV genome consists of two single-stranded (+)-sense RNA molecules, RNA1 and RNA2, which are encapsidated in a single virion (Krishna and Schneemann 1999) (Fig. 4). Both viral RNAs carry a methylated cap structure at the 5′ end but lack a poly(A) tail at the

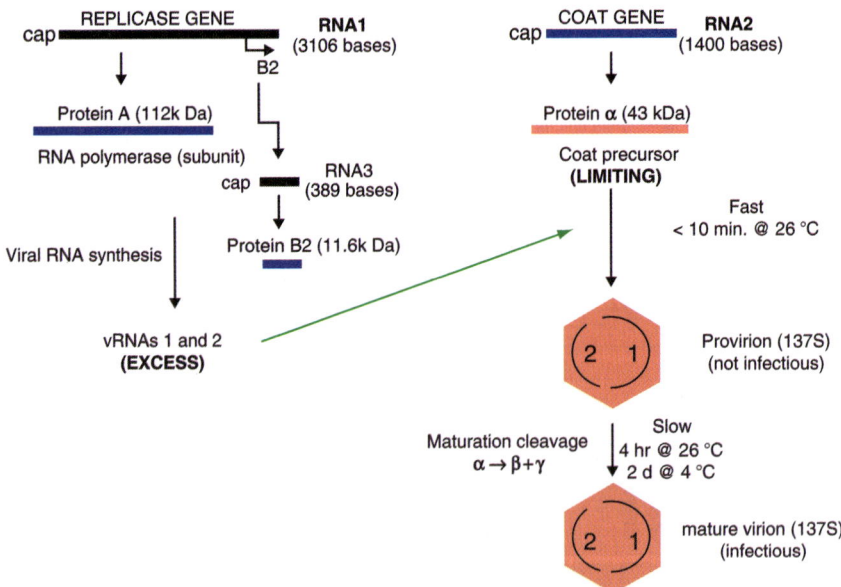

Fig. 4 Genomic organization and morphogenesis of flock house virus

3′ end (Dasgupta et al. 1984). RNA1 (3.1 kb) directs synthesis of protein A (112 kDa), the RNA-dependent RNA polymerase (Friesen and Rueckert 1981). RNA2 (1.4 kb) encodes protein alpha (43 kDa), the precursor of the coat protein (Friesen and Rueckert 1981). In addition to the genomic RNAs, infected cells contain a subgenomic RNA3 (0.4 kb) derived from the 3Î end of RNA1 (Friesen and Rueckert 1982). This RNA is not packaged into virions. It encodes protein B2 (11.6 kDa), an inhibitor of RNA silencing (Chao et al. 2005; Li et al. 2002).

In infected *Drosophila* cells, newly synthesized alpha protein first assembles into precursor particles called provirions (Gallagher and Rueckert 1988). Provirions contain 180 alpha subunits, arranged on a T=3 icosahedral lattice, and a copy of both RNA1 and RNA2. The assembly process triggers an autoproteolytic reaction in the 407 amino acid alpha chain, which results in cleavage between asparagine 363 and alanine 364 (Gallagher and Rueckert 1988). The newly formed polypeptides, beta (38 kDa, 363 amino acids) and gamma (5 kDa, 44 amino acids), remain associated with the mature virion. Maturation cleavage is required for acquisition of infectivity (Schneemann et al. 1992) and results in increased particle resistance to denaturing agents such as SDS and urea (Gallagher and Rueckert 1988).

The structure of FHV was solved by X-ray crystallography in the laboratory of Jack Johnson (Fisher and Johnson 1993). The 180 subunits of the FHV capsid can be divided into 60 triangular units, each of which consists of three chemically identical but conformationally slightly different protomers (Fig. 5a, b). (NB: a protomer refers to either coat precursor protein alpha or its cleavage products beta and gamma). The protomers are distinguished by the letter codes A, B and C and form the icosahedral asymmetric unit characteristic of a T=3 surface lattice. The three protomers form a prominent terraced pyramid (Fig. 5c, d). The central layer carries the bonding contacts between asymmetric units and forms the continuous part of the closed shell. The bottom layer is composed of helical segments that extend into the central cavity. This layer contains the beta-gamma cleavage site. The exterior surface consists of elaborate loops between the strands composing the beta barrel and these loops can be targeted for insertion of foreign peptides (Fig. 5c, d).

Propagation and Purification of Wild Type FHV and Strategies for Making Mutant Viruses

Wild type FHV grows exceptionally well in cultured *Drosophila* cells. From a 5- to 10-ml culture (corresponding to a single 100-mm culture dish), 1 mg of purified particles can be obtained. The purification protocol consists of a few simple centrifugation steps, mainly pelleting of virus in the cell lysate through a sucrose cushion and subsequent sedimentation through a continuous sucrose gradient. Further purification, for example by banding on CsCl gradients, is usually not required. The entire purification procedure can be easily completed within a day. To generate mutant FHV particles, two independent protocols have been developed

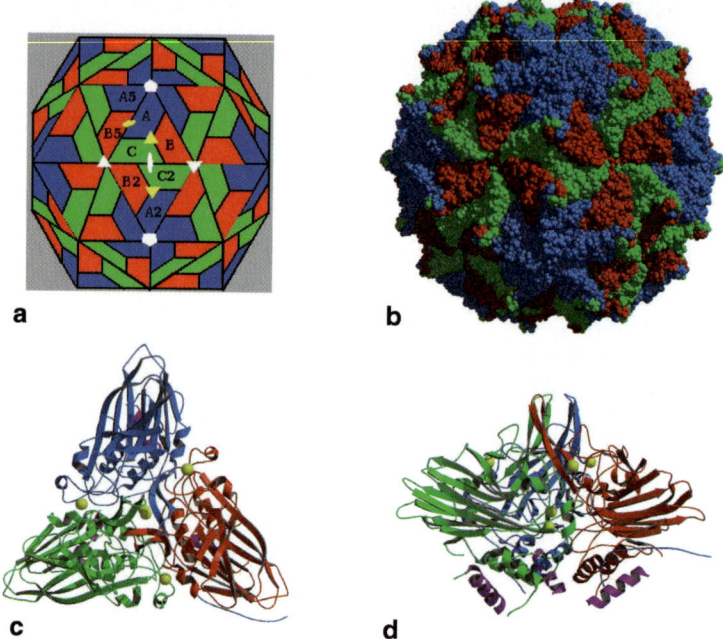

Fig. 5 a Schematic representation of the FHV capsid as a rhombic triacontahedron. Each trapezoid represents a protein subunit which consists of 407 amino acids. The labels *A*, *B* and *C* represent the three subunits in each of the 60 icosahedral asymmetric units in the T=3 particle. Although *A*, *B* and *C* represent chemically identical proteins, they are not related by strict symmetry and they are structurally slightly different. **b** Space-filling model of the FHV particle based on the X-ray structure at approximately 3-Å resolution. **c** Close-up view of the FHV asymmetric unit containing subunits A, B and C. View is from outside the virus down the quasi-threefold axis. Calcium ions located at the subunit interfaces are shown as *yellow spheres*. **d** Side view of the asymmetric unit showing loops present on the surface of the virion

(Fig. 6): transfection of *Drosophila* cells with *in* vitro synthesized FHV RNA transcripts or expression of the coat protein in *Spodoptera frugiperda-* (*Sf*21) cells or *Trichoplusia ni* cells via recombinant baculovirus vectors.

Transfection of *Drosophila* Cells

Liposome-mediated transfection of *Drosophila* cells with authentic nodaviral RNAs or transcripts synthesized in vitro from cDNA clones is a routine procedure in our laboratory (Schneemann and Marshall 1998). This protocol basically circumvents receptor-mediated endocytosis and release of RNA from whole virus particles. Once inside the cell, the naked RNAs are translated and the newly synthesized proteins initiate the infectious cycle. Progeny virions isolated from cells transfected with the wild type RNAs have been found to be identical in sedimentation rate and specific infectivity to those generated by infection of cells with virus particles (Schneemann et al. 1992). To generate mutant particles, amino acid changes such as insertions, deletions

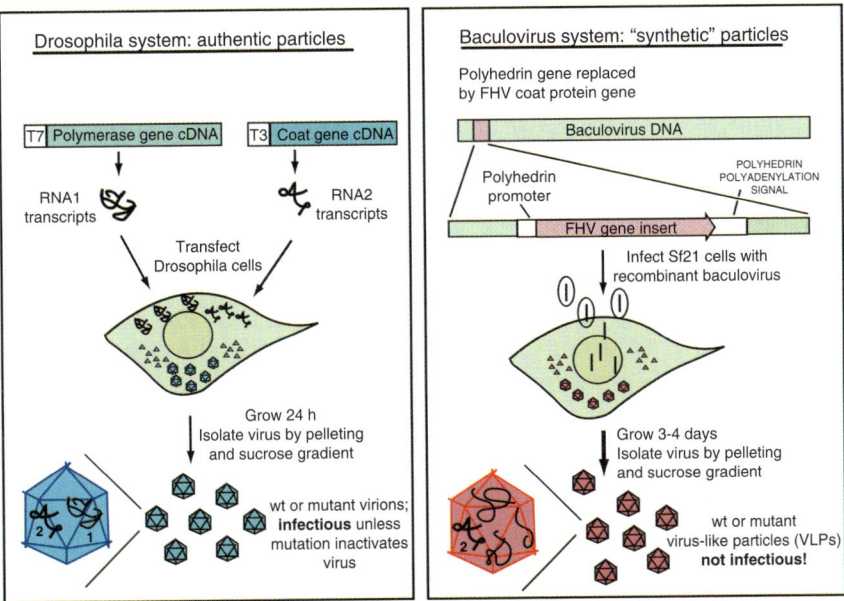

Fig. 6 A schematic diagram illustrating how FHV particles are generated by liposome-mediated transfection of *Drosophila* cells (*left panel*) or using recombinant baculovirus vectors (*right panel*)

or single site replacements are first introduced into the cDNA of RNA2 (coat protein gene) using standard procedures. In vitro synthesized mutant RNA2 transcripts are then transfected into *Drosophila* cells together with wt RNA1 (polymerase gene). The principal advantage of using transfected *Drosophila* cells to generate mutant FHV particles is the rapid generation time (24 h) and the ability to continue propagation of the mutants in the same cell line. However, there are also two significant caveats. First, the mutations introduced at the nucleotide level can interfere with or even inhibit replication of RNA2. This is primarily encountered when extensive deletions or insertions are introduced. Secondly, the mutation introduced at the protein level may block receptor binding of the particle and thereby inhibit infection of naïve cells. In the first scenario, it is not possible to obtain the mutant particles of interest, whereas in the second scenario the particle yield is restricted to the amount that is produced in the transfected cells without the benefit of secondary infections. This amount is typically in the microgram range.

Baculovirus Expression

Mutant or wt coat protein can also be expressed in *Sf*21 or *T. ni* cells using recombinant baculovirus vectors (Schneemann et al. 1993). In this case, the gene zamounts of protein alpha are synthesized in the late phase of baculovirus infection.

Expression of wild type coat protein in recombinant baculovirus-infected cells results in assembly of virus-like particles (VLPs) whose capsid structure is virtually indistinguishable from that of native particles, as demonstrated by X-ray crystallographic analysis. However, VLPs do not contain the normal complement of nodaviral RNAs 1 and 2 and thus cannot be infectious (Schneemann et al. 1993). Nodaviral RNA1 is not generated during expression of coat protein alpha in this system and RNA2 differs from native RNA2 owing to the presence of heterologous sequences at both its 5' and 3' ends. We have shown that the amount of RNA2 in a population of wt VLPs corresponds to 19% of the amount found in native virions (Krishna et al. 2002). The principal advantage of the baculovirus expression system is the fact that generation of particles is completely independent of FHV RNA replication and infectivity of the resulting capsids, thus circumventing both problems mentioned above. Moreover, the particle yield is exceptionally high. From a 1-l culture, 50–100 mg of purified VLPs have been obtained.

The FHV Chimera Technology: Insertion of Peptides and Protein Domains on the Surface of the Particle

As shown in Fig. 7, the FHV capsid protein contains a number of surface loops that can be targeted for the insertion of heterologous peptides and proteins. Four insertion sites located between amino acids 205 and 206 (site 1; most surface-exposed loop), amino acids 268 and 269 (site 2), 282 and 289 (site 3) and 136 and 137 (site 4) have been tested to date. Insertion of peptides ranging in size from 8–11 amino acids at sites 1 and 2 (Table 1) yielded homogeneous particles that could be produced in large quantities using a baculovirus expression system. Insertion at sites 3 and 4 has only yielded trace amounts of particles and attempts are underway to optimize the conditions for displaying foreign sequences at these locations.

It was also demonstrated that large domains derived from heterologous proteins can be successfully displayed on the surface of FHV at both sites 1 and 2. For example, a 137 amino acid Ig-like domain derived from the coat protein of the

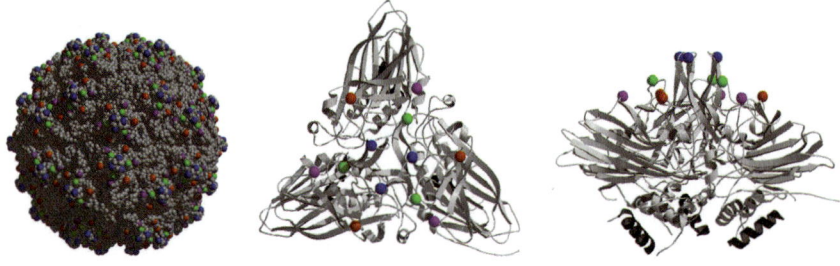

Fig. 7 Space-filling model of FHV capsid as well as top and side view of the asymmetric unit illustrating distribution of the four presentation sites on a single capsid. Site 1 (aa 205–206, *blue*); site 2 (aa 268–269, *green*); site 3 (aa 282–283, *red*), site 4 (aa 136–137, *purple*)

Table 1 Insertions made on the FHV capsid

Insertion	Site	Insertion size
FLAG epitope	1	8 amino acids
PA1 peptide	1, 2	11 amino acids
Ig-like domain from NωV	1	137 amino acids
ANTXR2 I-domain	1, 2	181 amino acids

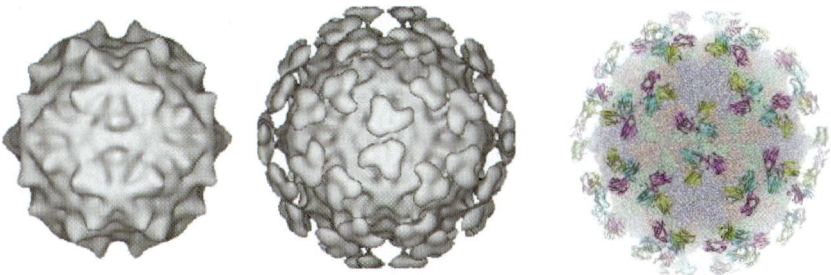

Fig. 8 *Left and center panel* Three-dimensional, surface-shaded reconstruction of wt FHV and FHV-Ig chimeric particles, respectively. Note the additional densities at the quasi-threefold axis of FHV-Ig not observed in wt FHV. *Right panel* Pseudoatomic model of FHV chimeric particles showing approximate orientation of the 180 Ig-like domains on the surface of the virion

insect virus Nudaurelia w capensis virus was inserted at site 1. The chimeric particles were stable, homogeneous and could be generated in large amounts. Structural analysis of the particles by cryoEM and image reconstruction demonstrated that the chimeric particles contained additional density at the expected locations and a pseudo atomic model showed the approximate orientation of the Ig-like domains on the surface of FHV (Fig. 8). Based on the structural data, the Ig-like domain appeared to be folded correctly, a notion that was confirmed by the fact that the chimeric particles could be precipitated with protein G. The 181 amino acid extracellular domain of capillary morphogenesis protein-2, a receptor for anthrax toxin, was also successfully inserted at sites 1 and 2 (Manayani et al. 2007). The insertion was well tolerated and structural analysis demonstrated that the protein fold was similar if not identical to that observed in the native protein. Moreover, the chimeric particles behaved as competitive inhibitors of anthrax toxin in a cell intoxication assay (see Sect. 3.2.3). Taken together, these results indicate that foreign protein domains, particularly those that are useful for tumor targeting, can be displayed on the surface of FHV.

Chemical Labeling Studies of FHV Amino Acid Replacement Mutants

FHV particles containing appropriate amino acid replacements can be conjugated to fluorescent dyes using straightforward bioconjugation methods. Specifically, mutants of FHV displaying lysine or cysteine residues at sites 1–4 (e.g., A205K, A282K, S268K, S136K, A282C, S268C and S136C) can be conjugated to fluorescein using fluorescein-isothiocyanate (FITC), fluorescein-N-hydroxyl-succinimidoester and fluorescein-maleimide (Fig. 9). The first two compounds react with free amino groups (i.e., lysine residues), whereas the third compound reacts with sulfhydryl groups (i.e., cysteine residues). Wild type FHV particles do not yield conjugates under identical conditions because they do not contain amino or sulfhydryl groups exposed on their surface. The mutant particles are prepared by transfection of *Drosophila* cells with wt FHV RNA1 and mutated RNA2. The resulting mutant particles are viable and can be grown to high titers in the *Drosophila* cell line. In general, 60–100 coat protein subunits of particles containing an exposed lysine residue can be chemically modified with the compounds indicated above, while all 180 subunits can be modified with the fluorescein-maleimide. Thus, the reactivity of the exposed cysteine residues is somewhat higher under the conditions tested so far. Double mutants that contain both reactive lysines and cysteines on the surface have also been generated to modify the particles with different compounds using sequential reactions schemes.

The simple chemistries developed for CPMV and FHV attachments allow a plug-in format that allows the potential for further display of molecules in a spatially controlled environment that should facilitate the most optimal function of ligand–receptor interactions for biomedical interactions. It has been shown that CPMV in particular, and to a lesser extent FHV, are stable to a variety of organic solvents and reaction conditions (Portney et al. 2005; Wang et al. 2002a, 2002b, 2002c). The chemistry has made it possible to significantly expand the range and type of molecules that can be displayed on CPMV or FHV (Singh et al. 2005). Portney et al.

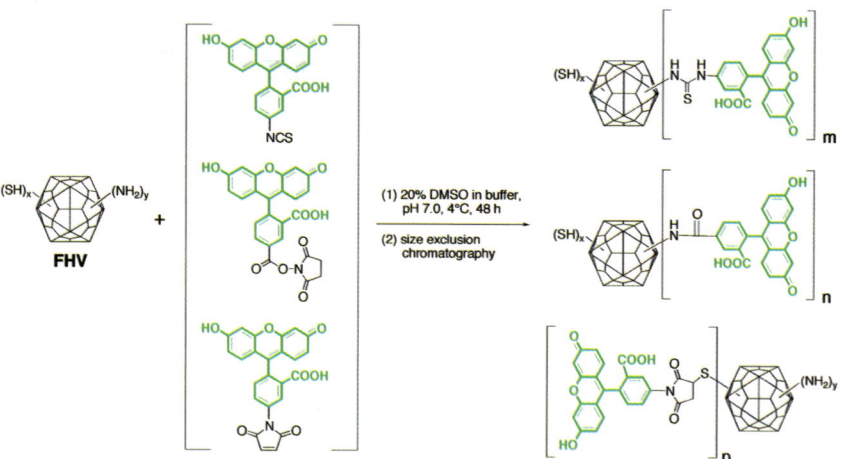

Fig. 9 Schematic diagram illustrating the chemistry performed on FHV. All three compounds are derivatives of fluorescein. The top two compounds, flurescein-isothiocyanate (FITC) and fluorescein-N-hydroxyl-succinimidoester, react with free amino groups on the surface of FHV, whereas the bottom compound, fluorescein-maleimide, reacts with free sulfhydryl groups

also describe the use of CPMV and FHV for materials applications such as ordering of networks containing viruses and QD or single-walled carbon nanotubes (SWCNT) (Portney et al. 2005). Other types of 2D arrays have resulted from cross-linking single-stranded DNA on the surface of the virus and creating an array by hybridization with other particles displaying the complementary nucleic acid sequence (Strable et al. 2004). Further, it is possible using CPMV to array metals in three dimensions is found in the work of Wang et al., where it was shown that Nanogold could decorate the CPMV surface at specifically engineered cysteine contact points (Wang et al. 2002b). Finally much progress has been made conjugation of tumor ligands onto the CPMV surface, as described in Sect. 3 (Chatterji et al. 2004b; Destito 2007; Sen Gupta et al. 2005a).

Applications of CPMV and FHV Particles to Therapeutics and Tumor Targeting

Tumor Targeting

The natural container-like properties of viruses have been attractive for gene delivery for many years, and are now being harnessed for therapeutic delivery and nanotechnology. Virus-based nanoparticles genetically modified to display targeting peptides together with an imaging agent or drug loading moieties appears to be a logical choice for targeted delivery as well as an attractive strategy to target cancer. Combinatorial library methods have proved to be a pivotal tool in identification of molecules of interest for achieving therapeutic targeting. Typically such libraries take advantage of phage display techniques, whereby it is possible to screen large libraries of short peptides for their specific affinity for cell-surface ligands. Generation of a combinatorial phage display library is achieved by genetic modification of the bacteriophage (typically M13 or lambda), displaying peptides on the virus surface followed by screening and identification of peptides in the interacting phages. Peptides may be of clinical relevance in tumors that express their respective receptors in high amounts.

Vascular Imaging Using Wild Type CPMV

A recent study to evaluate the ability of CPMV to visualize the vasculature showed that fluorescently labeled CPMV particles are extremely useful for intravital vascular imaging in live mouse and chick embryos (Lewis et al. 2006). In this study, CPMV gave images superior to many commonly used particles such as dextrans, fluorescent microspheres, and lectins. Furthermore, the particles appeared to have a natural affinity for vascular endothelium, particularly labeling the venous vasculature. Such targeting appears to be due to interaction with a cell-surface protein that is currently being characterized (Koudelka et al. 2007). These studies indicate that natural and targeted VNPs will be extremely useful tools for visualizing normal and diseased vasculature such as in growing tumors (Lewis et al. 2006).

Targeting CPMV and FHV to Tumors

Efforts to adapt icosahedral virus particles for use as platforms for the multivalent presentation of cell-surface ligands have been described recently (Douglas and Young 1999; Wang et al. 2002c). One example under investigation in our laboratory is the specific targeting of CPMV and FHV nanoparticles to neuroblastoma tumor cells using specific peptides. The neuropeptide Y (NPY) analog was previously identified by screening several high-affinity binding peptides to the Y_1 receptors that are overexpressed on human neuroblastoma SK-N-MC cells (Soll et al. 2001). NPY is a neurotransmitter that belongs to the family of pancreatic polypeptides. These polypeptides are important modulators of the central and peripheral nervous system and NPY is one of the most abundant neuropeptides in the brain. The receptors for NPY are produced in a number of neuroblastoma and similarly derived cell lines. So far, five receptor subtypes (Y1, Y2, Y4, Y5, Y6) have been identified that bind NPY with nanomolar affinity, providing a convenient way to specifically target the neuroblastoma cells by receptor-mediated endocytosis (Michel 1998). SK-N-MC tumor cells selectively express the Y_1 receptor (Gordon et al. 1990; Larhammar 1992), and it has been shown that a 36-residue NPY analog (Phe7-Pro34) can specifically bind NPY-Y_1 that is expressed on human neuroblastoma tumors (Soll et al. 2001).

In a recent study, we have shown the ability of chimeric viruses to be used as targeting agents for cancer detection, using CPMV virus genetically modified to display the NPY peptide as targeting ligand. In these experiments, the 36-residue NPY peptide was displayed on CPMV in the βB-βC loop. Although this was a relatively large insertion for CPMV, the virus was able to replicate at sufficient levels to characterize the virus. Studies to evaluate the CPMV-NPY binding specificity showed that it interacted specifically with SK-N-MC cells expressing the Y1 receptor (Fig. 10).

Subsequent studies to display the NPY peptide on the surface of FHV have found that the VLP replication system of FHV is somewhat more amenable to display of larger insertions such as NPY. In these studies, the same NPY insertion was introduced at the 205 site on FHV. In this system, the VLPs assembled and were produced at high levels, similar to wild type FHV-VLPs. Characterization of the specificity of FHV-NPY particles for neuroblastoma cells in vitro and in vivo is currently underway.

In a separate study, we chose a single-chain antibody recognizing the carcinoembryonic antigen (CEA) as our tumor-targeting motif. Carcinoembryonic antigen (CEA) is a highly characterized, cell-surface glycoprotein specifically overexpressed by a variety of tumor cells, which provides a tool for tumor-specific recognition (Goldenberg 1992; Hammarstrom et al. 1989). The single-chain (scFv) CEA

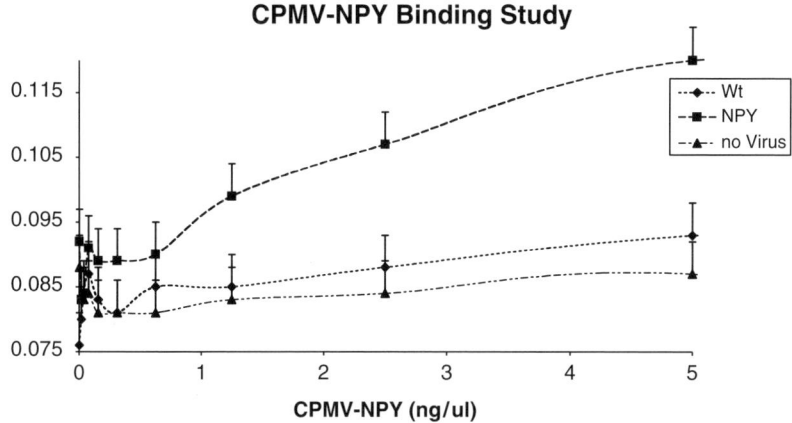

Fig. 10 Binding of CPMV-NPY to SK-N-MC neuroblastoma cells

antibody is well characterized and recognizes CEA on a wide variety of tumor tissues while it does not bind to normal tissues. Thus the anti-CEA strategy provides an opportunity to determine whether it is possible to target CPMV particles specifically to CEA-expressing cells in vitro and in vivo. Recently, we showed that it was possible to append chemical groups to the surface of CPMV by conjugation to exposed lysines that the crystal structure showed are found on the capsid surface and to cysteines installed by mutagenesis (Fig. 11). Such processes have been used to attach several proteins to the viral capsid, including scFv CEA antibody and transferrin (Chatterji et al. 2002, 2004b; Sen Gupta et al. 2005a).

CEA antibody (anti-CEA) was expressed in bacteria using a 700-nucleotide cDNA encoding the single-chain Ab under the control of the T7 transcription promoter. This construct contains the immunoglobulin heavy and light chains of the antibody attached via a flexible linker. The antibody sequence, together with a myc epitope sequence (for detection) and a cysteine residue was introduced at the C-terminus of the antibody to allow chemical attachment of the anti-CEA to CPMV (Fig. 11A). Chemical conjugation of the scFc to CPMV was carried out using sulfo-SMCC as bifunctional linker. This linker contains a NHS-ester group that reacts with the lysines of the CPMV capsid and a maleimide group that is selective for the cysteine at C-terminus of scFv (Fig. 11). Binding of antibody-coupled CPMV to CEA-expressing cells in vitro in comparison to wild type CPMV showed that CPMV-Ab bound specifically to human tumor cells expressing CEA (HT-29), while wild type CPMV did not. Further, in a murine-human tumor xenograft model, CPMV-Ab inoculated intravenously confirmed homing of CPMV-Ab in the tumors (Fig. 11d, e). These preliminary data suggest that it is possible to target CPMV particles specifically to tumor cells in vivo.

Finally, in collaboration with the Finn laboratory, our group recently showed that using azide-alkyne click chemistry it was possible to efficiently attach several ligands to CPMV. First, the tumor ligand folic acid (FA) was attached in variable

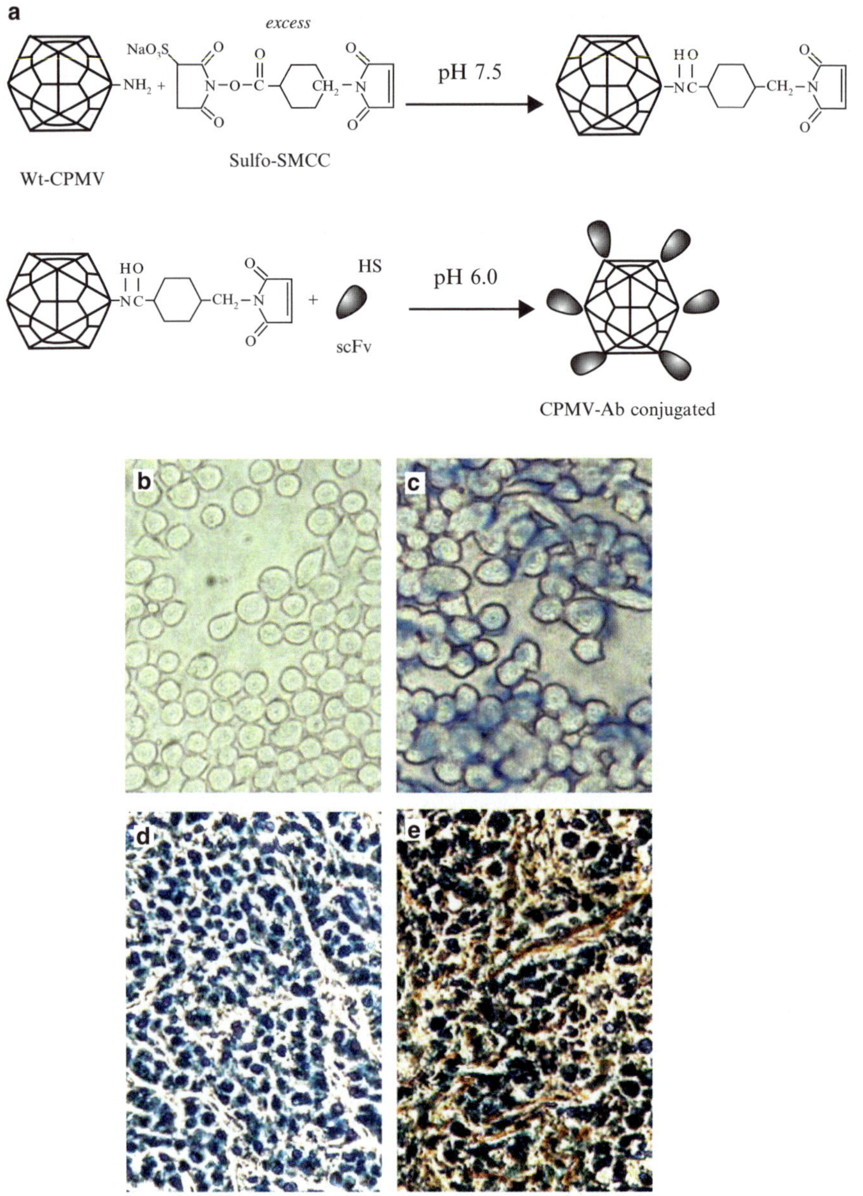

Fig. 11 a Chemical conjugation of CPMV to scFv anti-CEA. **b** HT-29 cells incubated with wild type CPMV, or **c** CPMV-CEA. **d** HT-29 tumor section showing localization of wild type CPMV, **e** localization of CPMV-CEA virus in the tumor

numbers using click chemistry, and this attachment facilitated specific cell targeting (Destito 2007). Second, the very large (80-kD) transferrin (Tfn) molecule was conjugated to the surface of CPMV (Sen Gupta et al. 2005a). This is the largest exogenous protein attached to CPMV to date (Chatterji et al. 2004b). Our studies showed that using this highly efficient chemical reaction it was possible to attach Tfn such that it maintained its affinity for Tfn receptor. This technology allows the multivalent display of much larger ligands and should facilitate more complex modes of tumor and other cell-specific targeting strategies using viral nanoparticles.

Vaccine Applications for CPMV and FHV

Both CPMV and FHV nanoparticles have been developed for vaccine applications. Two of the main reasons for exploiting the CPMV technology in particular for vaccine applications are the possibility for an edible vaccine strategy, and for the inherent stimulation of responses at mucosal surfaces, particularly via the intranasal or oral routes.

Antibody Responses Induced by Peptide Antigens Displayed on CPMV

Since it has a highly ordered capsid structure, CPMV was predicted to be an efficient inducer of antibodies without requiring additional adjuvants (Durrani et al. 1998). Specific antibody responses against a variety of epitopes have been demonstrated, including the Nlm1A epitope from human rhinovirus, the 731–752 epitope from HIV envelope glycoprotein, and the VP2 protein of mink enteritis virus (Dalsgaard et al. 1997; Durrani et al. 1998; Taylor et al. 2000). For bacterial antigens, it has been shown that the *Staphylococcus aureus* fibronectin binding protein epitope, when presented on CPMV, raises antibody responses that are protective against challenge in rats (Rennermalm et al. 2001). In addition, CPMVs displaying *Pseudomonas* antigens have been developed (Gilleland et al. 2000). Most recently, CPMV has been used for anti-malaria applications as well as for driving the production of heterologous vaccine antigens in plants (Canizares et al. 2005; Liu et al. 2005; Mechtcheriakova et al. 2005; Yasawardene et al. 2003).

VNPs as Inducers of Cellular Immune Responses: Comparison with Other VLP Systems

Since in the CPMV system infectious viruses are used for presentation, they are termed viral nanoparticles (VNPs). Similar approaches have been utilized with VLP systems, whereby virus capsid/coat proteins are expressed in the absence of viral RNA and assemble into noninfectious particles that can be purified. Some examples of VLP systems for display of T cell epitope peptides for vaccine development are papillomavirus, parvovirus, and cores of hepatitis B virus (Da Silva et al. 2003; Martinez et al. 2003; Sedlik et al. 1997). In general, these VLP systems utilize heterologous expression strategies, typically baculovirus expression in insect cells or in yeast, to produce the VLPs; however, as described in Fig. 6, VLPs cannot replicate on their own.

Cellular immune responses are key to protection from many viral diseases as well as some bacterial diseases. While live-attenuated vaccines are typically more efficient at inducing a cytotoxic T lymphocyte (CTL) response, most live-attenuated vaccines are expensive to produce, unstable to tropical temperatures, and are often delivered by injection. The benefits of utilizing a nonreplicating antigen for safety and stability to transport make it extremely worthwhile to develop epitope delivery systems that can induce protective CTL as well as humoral immunity while maintaining these benefits.

The mechanisms by which antigen-presenting cells (APCs) can take up exogenous antigens such as VLPs and present them in the context of MHCI is termed cross-presentation (Bevan 1976; Heath and Carbone 2001; Yewdell et al. 1999b). This phenomenon contradicts the classical models of antigen presentation whereby MHC-I presentation is dependent on endogenous synthesis of antigens presented by MHC-I. Nevertheless, the results with a variety of systems, including VLP vaccination, clearly indicate that presentation of antigens in the absence of endogenous protein synthesis in professional APCs such as DC is possible and is a viable tool for vaccine development (Yewdell et al. 1999a). There has been considerable controversy in the field regarding whether cross-presentation is the normal route for presentation of viral antigens that do not naturally infect APCs (Zinkernagel 2002). Nevertheless, the fact that cross-presentation does occur and can be harnessed as a technique to induce cellular responses against nonreplicating antigens suggests that it is an important and viable vaccination development strategy, which has the potential to combine the benefits of safety, deliverability, and simultaneous antibody and CTL induction.

Taking advantage of the cross-presentation phenomenon, induction of CTL has been achieved in the papillomavirus and parvovirus VLP systems, suggesting that uptake of VLPs or VNPs by antigen-presenting cells, followed by presentation of the epitopes by MHC class I, is possible using these systems (Da Silva et al. 2003). Indeed, with the parvovirus VLP system, CTLs can be detected ex vivo and animals are protected against subsequent pathogen challenge (Martinez et al. 2003; Sedlik et al. 1997). Plant viruses have only recently been explored for inducing CTL activity in vivo. Our preliminary studies indicate that APCs including CD11c$^+$/CD8a dendritic cells readily internalize CPMV both in vitro and in vivo. It will be interesting to determine whether this localization results in cross-presentation or whether further intracellular targeting is required (M. Gonzalez, C. Rae, M. Manchester, unpublished data).

FHV-Based Vaccines and Epitope Presentation Systems

Vaccine strategies utilizing FHV are less widely used than for CPMV; however, promising results have been obtained in several areas, including induction of humoral immunity and using multivalent FHV particles as a display system for detecting and quantifying antibodies in immune sera. For example, several studies have shown that portions of the HIV-1 gp120 V3 loop, or a fragment of gp41, can be presented on FHV and induce peptide-specific immune responses in mice (Buratti et al. 1996; Scodeller et al. 1995). This epitope display strategy is particularly useful for presenting conformational epitopes that are difficult to use in a soluble peptide format for priming an antibody response, similar to what was observed for the conformational Nlm1A epitope, which was preferentially recognized by antibodies when displayed in a conformationally constrained manner on CPMV (Taylor et al. 2000). A second focus has been on using immunoreactive peptides presented multivalently as fusions with the FHV coat protein, as a means to screen for the presence of epitope-specific antibodies in human sera, using hepatitis C virus or HIV-1 gp120 or gp41 (Buratti et al. 1997, 1998; Schiappacassi et al. 1997). Finally, nodavirus particles themselves have proven to be potent immunogens in their host species: an example is immunization of fish with a nodavirus VLP offering protection against subsequent infection (Thiery et al. 2006).

The demonstrated capacity for multivalent display of large or more conformationally complex structures on the FHV surface compared to CPMV suggested that FHV particles may be more amenable for vaccine design in the long term. As mentioned above, it was recently shown that it was possible to display the Ig-like I-domain from the human anthrax toxin receptor ANTXR2 on the surface of FHV. This display led to antitoxin activity (see Sect. 3.3), but also the particles could act as a scaffold for display of the ANTXR2 ligand, the protective antigen (PA) of *Bacillus anthracis*. Immunization with FHV particles displaying PA in this multivalent manner led to a fast-acting immunogen that could induce neutralizing antibody responses against PA in only 3 weeks. These antibodies in turn could protect animals against a lethal challenge of anthrax toxin after the single immunization (Manayani et al. 2007). These results indicate that FHV may be an important platform for further display of microbial antigens.

Virus-Based Therapeutics: Antivirals and Antitoxins

We developed the first CPMV-based antiviral therapeutic, a nanoparticle designed to inhibit a virus–cellular receptor interaction by multivalent display of a fragment of the receptor on CPMV. Our initial report of the CPMV-antiviral inhibited infection efficiently in tissue culture where it was approximately 20- to 100-fold more potent than an equimolar concentration of free peptide, while the wild type CPMV virus had no inhibitory effect on infectivity (Khor et al. 2002). Our more recent estimates suggest between two- and 20-fold enhancement over the monovalent format. Our study was the first to show that multivalent presentation on the CPMV nanoparticle surface retained the function of monovalent materials and was the first example of a nanoparticle inhibitor of ligand–receptor interactions. More recently, experiments with multivalent carbohydrates agglutinating lectins on CPMV show a multivalent effect of approximately 180-fold over free carbohydrates (Sen Gupta et al. 2005b), further demonstrating that multivalent display of ligands on CPMV is a potentially effective way to improve the binding of a ligand of interest by between one and two orders of magnitude.

Studies with the FHV-ANTXR2 particles described above also showed inhibitory antitoxin activity against *B. anthracis* lethal toxin. However, in this case a very mild multivalent effect was observed (less than twofold) and this effect was observed only in vivo (Manayani et al. 2007). Thus the generation of a multivalent effect is not always observed in VNP-based systems and is probably dependent upon the particular ligand–receptor pair that is being targeted.

Advantages and Disadvantages Comparing CPMV and FHV Type Systems

When comparing the two systems, each has advantages for particular uses, in particular the ease of production of wild type and mutant viruses at high yields, the ability to introduce complex genetic modifications into the viruses, and chemical stability. Wild type CPMV and some mutants are produced at very high yields in plants, and this production is extremely inexpensive. However, when more complex mutations are introduced on the capsid surface (in our experience greater than 25–30 amino acids), the yield is not as high. Similarly, if the mutation introduced interferes with infectivity in plants the yield will be low to nonexistent. Because of the VLP expression system available for FHV, it is not necessary to rely on infectious viruses, and it is therefore possible to introduce more complex insertions and mutations on the capsid surface. Several examples are the insertions of protein domains at up to 150 amino acids that have been performed in the Schneemann laboratory (Manayani et al. 2007).

Chemical modification of CPMV has been more thoroughly characterized than for FHV, and in the limited comparisons that have been performed, CPMV is more stable to withstand harsh chemical reaction conditions. CPMV also has more natural attachment points via surface lysine residues, in comparison to wild type FHV that has none. However, mutations to introduce these residues on FHV have been straightforward and simply require an additional step when making other targeting mutations to include a background of chemically modifiable capsid when appropriate.

The Future of Viral Nanoparticles for in Vivo Therapeutic and Diagnostic Purposes

Since we are interested in developing VNPs as multivalent tools for targeting and imaging in vivo, it has been important to first determine the natural bioavailability, immunogenicity, and toxicity of VNPs.

Evaluating the Biodistribution of VNPs in Vivo

Bioavailability of CPMV and FHV

We first studied the delivery of CPMV nanoparticles including oral and intravenous routes (Rae et al. 2005). We inoculated animals by each route with purified CPMV particles followed by detection of the viral particles in vivo (Rae et al. 2005). Single-stranded viral RNA is naturally encapsidated inside CPMV particles and became our first tool for detecting the trafficking of particles. Mice were fed wild type CPMV via oral gavage and tissues isolated. CPMV RNA could be detected by RT-PCR in the spleen, kidney, liver, gastrointestinal tract, blood, lungs, lymph nodes, brain, bone marrow, and salivary gland. Similar results were obtained using fluorescently labeled CPMV as well. These studies suggest that CPMV can enter the bloodstream and lymphoid systems following oral administration and can access a variety of tissues in the animal. Similar results were also obtained when animals were inoculated with CPMV intravenously followed by perfusion prior to harvesting the tissues. We also conducted experiments to show that ingestion of virus-infected leaves also led to a similar trafficking pattern, indicating that the in vivo trafficking is not an artifact of the gavage procedure. It is important to note that virus replication in the animal is highly unlikely because CPMV does not replicate in animal cells. Rather, the CPMV likely travels through the gut and into the tissues, as CPMV labeled with AF488 is readily observed in Peyer's patch (PP) epithelial tissue in the ileum region of the mouse intestine. PPs are responsible for particulate uptake in the intestine (C. Rae, M. Gonzalez, and M. Manchester, unpublished data). In addition, trafficking studies conducted in the presence of a preexisting antibody response showed that such a response did not impede trafficking, as has been observed in other systems (M. Gonzalez, C. Rae, and M. Manchester, unpublished data; Mandl et al. 2001; Ruedl et al. 2005). Together our studies indicate that CPMV particles are deliverable via the intravenous or oral routes. These results suggest that CPMV has potential to be a highly useful bioavailable nanoparticle platform for in vivo studies.

Similar studies have been initiated using FHV administration by the oral route (P. Singh, A. Schneemann, unpublished data). Similar to CPMV, FHV could be observed in the bloodstream and in a variety of tissues following oral feeding. These results suggest that the size and stability of the particles, rather than their particular surface characteristics, govern their trafficking across the intestinal epithelium.

Immunologic Properties of CPMV in Vivo

To evaluate the antibody response against VNPs, the serum IgG responses in mice that had been immunized with CPMV or CPMV particles decorated with polyethylene glycol and fluorescein CPMV-PEG-F have been evaluated (Raja et al. 2003). The immune response was analyzed using ELISA to detect CPMV specific antibodies. A serum antibody response was generated against CPMV, and this response was significantly reduced or eliminated with the modified polymer forms. These results are consistent with our studies of attachment to endothelial cells in vivo being blocked by PEG coating (Lewis et al. 2006).

Toxicity Studies of CPMV in Vivo

To address the future feasibility of using VNPs in vivo, the potential toxicity has also been evaluated. To date, studies in the Manchester laboratory have investigated the maximum tolerated dose (MTD) of CPMV in animals using intravenous doses of up to 100 mg/kg of purified virus (Singh et al. 2007). Necropsy and histologic examination of tissues from all major organ systems showed no evidence of toxicity. Mice did not show any clinical signs that were different from saline-injected mice at all of the time points observed. This lack of toxicity contrasts favorably with other viral particles used in vivo such as adenovirus particles that typically show high liver toxicity at doses of 5×10^9 particles/mouse (Engler et al. 2004), where with CPMV doses of up to 10^{14} particles do not result in toxicity (Singh et al. 2007). Studies of doses higher than 100 mg/kg are in progress.

Conclusions

The use of viruses as nanoparticles for biomedical applications is still in the early stages. However, the properties of structurally characterized viruses are extremely well-suited for nanotechnology. The combination of a defined, multivalent structure that can be tailored for specific interactions and targeting in vivo, along with the natural container-like shape of virus particles, makes viruses highly appropriate tools for further development.

References

Bevan MJ (1976) Cross-priming for a secondary cytotoxic response to minor H antigens with H-2 congenic cells which do not cross-react in the cytotoxic assay. J Exp Med 143:1283

Brennan FR, Jones TD, Hamilton WD (2001) Cowpea mosaic virus as a vaccine carrier of heterologous antigens. Mol Biotechnol 17:15–26

Brumfield S, Willits D, Tang L, Johnson JE, Douglas T, Young M (2004) Heterologous expression of the modified coat protein of Cowpea chlorotic mottle bromovirus results in the assembly of protein cages with altered architectures and function. J Gen Virol 85:1049–1053

Buratti E, Tisminetzky SG, Scodeller ES, Baralle FE (1996) Conformational display of two neutralizing epitopes of HIV-1 gp41 on the Flock House virus capsid protein. J Immunol Methods 197:7–18

Buratti E, Di Michele M, Song P, Monti-Bragadin C, Scodeller EA, Baralle FE, Tisminetzky SG (1997) Improved reactivity of hepatitis C virus core protein epitopes in a conformational antigen-presenting system. Clin Diagn Lab Immunol 4:117–121

Buratti E, McLain L, Tisminetzky S, Cleveland SM, Dimmock NJ, Baralle FE (1998) The neutralizing antibody response against a conserved region of human immunodeficiency virus type 1 gp41 (amino acid residues 731–752) is uniquely directed against a conformational epitope. J Gen Virol 79:2709–2716

Canizares MC, Lomonossoff GP, Nicholson L (2005) Development of cowpea mosaic virus-based vectors for the production of vaccines in plants. Expert Rev Vaccines 4:687–697

Chao JA, Lee JH, Chapados BR, Debler EW, Schneemann A, Williamson JR (2005) Dual modes of RNA-silencing suppression by Flock House virus protein B2. Nat Struct Mol Biol 12:952–957

Chatterji A, Burns LL, Taylor SS, Lomonossoff G, Johnson JE, Lin T, Porta C (2002) Cowpea mosaic virus: from the presentation of antigenic peptides to the display of active biomaterials. Intervirology 45:362–370

Chatterji A, Ochoa W, Paine M, Ratna BR, Johnson JE, Lin T (2004a) New addresses on an addressable virus nanoblock: uniquely reactive lys residues on cowpea mosaic virus. Chem Biol 11:855–863

Chatterji A, Ochoa W, Shamieh L, Salakian SP, Wong SM, Clingon G, Ghosh P, Lint T, Johnson J (2004b) Chemical conjugation of heterologous proteins on the surface of cowpea mosaic virus. Bioconjug Chem 15:807–813

Da Silva DM, Schiller JT, Kast M (2003) Heterologous boosting increases immunogenicity of chimeric papillomavirus virus-like particle vaccines. Vaccine 21:3219–3227

Dalsgaard K, Uttenthal A, Jones TD, Xu F, Merryweather A, Hamilton W, Langeveld J, Boshuizen R, Kamstrup S, Lomonossoff G, Porta C, Vela C, Casal J, Meloen R, Rodgers P (1997) Plant-derived vaccine protects target animals against a viral disease. Nat Biotechnol 15:248–252

Dasgupta R, Ghosh A, Dasmahapatra B, Guarino LA, Kaesberg P (1984) Primary and secondary structure of black beetle virus RNA2, the genomic messenger for BBV coat protein precursor. Nucleic Acids Res 12:7215–7223

Destito G Yeh R, Rae CS, Finn MG, Manchester M (2007) Folic acid-mediated targeting of cowpea mosaic virus particles to tumor cells. Chem Biol 14:1152–1162

Douglas T, Young MJ (1998) Host-guest encapsulation of materials by assembled virus protein cages. Nature 393:152–155

Douglas T, Young M (1999) Virus particles as templates for materials synthesis. Adv Mater 11:679–681

Durrani Z, McInerney TL, McLain L, Jones T, Bellaby T, Brennan FR, Dimmock NJ (1998) Intranasal immunization with a plant virus expressing a peptide from HIV-1 gp41 stimulates better mucosal and systemic HIV-1-specific IgA and IgG than oral immunization. J Immunol Meth 220:93–103

Engler H, Machemer T, Philopena J, Wen SF, Quijano E, Ramachandra M, Tsai V, Ralston R (2004) Acute hepatoxicity of oncolytic adenoviruses in mouse models is associated with expression of wild-type E1a and induction of TNF-alpha. Virology 328:52–61

Ferrari M (2005) Cancer nanotechnology: opportunities and challenges. Nat Rev Cancer 5:161–171

Flenniken ML, Willits D, Harmsen AL, Liepold LO, Harmsen AJ, Young MJ, Douglas T (2006) Melanoma and lymphocyte cell-specific targeting incorporated into a heat shock protein cage architecture. Chem Biol 13:161–170

Friesen PD, Rueckert RR (1981) Synthesis of black beetle virus proteins in cultured Drosophila cells: differential expression of RNAs 1 and 2. J Virol 37:876–886

Friesen PD, Rueckert RR (1982) Black beetle virus: messenger for protein B is a subgenomic viral RNA. J Virol 42:986–995

Gallagher T, Rueckert RR (1988) Assembly-dependent maturation cleavage in provirions of a small icosahedral insect ribovirus. J Virol 62:3399–3406

Gilleland HE, Gilleland LB, Staczek J, Harty RN, Garcia-Sastre A, Palese P, Brennan FR, Hamilton WD, Bendahmane M, Beachy RN (2000) Chimeric animal and plant viruses expressing epitopes of outer membrane protein F as a combined vaccine against *Pseudomonas aeruginosa* lung infection. FEMS Immunol Med Microbiol 27:291–297

Goldenberg DM (1992) Cancer imaging with CEA antibodies: historical and current perspectives. Int J Biol Markers 7:183–188

Gordon EA, Kohout TA, Fishman PH (1990) Characterization of functional neuropeptide-Y receptors in a neuroblastoma cell-line. J Neurochemistry 55:506–513

Hammarstrom S, Shively JE, Paxton RJ, Beatty BG, Larsson A, Ghosh R, Bormer O, Buchegger F, Mach JP, Burtin P et al (1989) Antigenic sites in carcinoembryonic antigen. Cancer Res 49:4852–4858

Harisinghani MG, Barentsz J, Hahn PF, Deserno WM, Tabatabaei S, van de Kaa CH, de la Rosette J, Weissleder R (2003) Noninvasive detection of clinically occult lymph-node metastases in prostate cancer. N Engl J Med 348:2491–2499

Heath WR, Carbone FR (2001) Cross-presentation, dendritic cells, tolerance and immunity. Ann Rev Immunol 19:47

Hooker JM, Kovacs EW, Francis MB (2004) Interior surface modification of bacteriophage MS2. J Am Chem Soc 126:3718–3719

Johnson J, Lin T, Lomonossoff G (1997) Presentation of heterologous peptides on plant viruses: genetics, structure, and function. Annu Rev Phytopathol 35:67–86

Khor IW, Lin T, Langedijk JP, Johnson JE, Manchester M (2002) Novel strategy for inhibiting viral entry by use of a cellular receptor-plant virus chimera. J Virol 76:4412–4419

Kickhoefer VA, Garcia Y, Mikyas Y, Johansson E, Zhou JC, Raval-Fernandes S, Minoofar P, Zink JI, Dunn B, Stewart PL, Rome LH (2005) Engineering of vault nanocapsules with enzymatic and fluorescent properties. Proc Natl Acad Sci U S A 102:4348–4352

Klem MT, Willits D, Young M, Douglas T (2003) 2-D array formation of genetically engineered viral cages on au surfaces and imaging by atomic force microscopy. J Am Chem Soc 125:10806–10807

Koudelka K, Rae CS, Gonzalez MJ, Manchester M (2007) Interaction between a 54-kilodalton mammalian cell surface protein and cowpea mosaic virus. J Virol 81:1632–1640

Krishna NK, Schneemann A (1999) Formation of an RNA heterodimer upon heating of nodavirus particles. J Virol 73:1699–1703

Larhammar D, Blomqvist AG, Yee F, Jazin E, Yoo H, Wahlested C (1992) Cloning and functional expression of a human neuropeptide Y/peptide YY receptor of the Y1 type. J Biol Chem 267:10935–10938

Lee SW, Mao C, Flynn CE, Belcher AM (2002) Ordering of quantum dots using genetically engineered viruses. Science 296:892–895

Levy LS, Yudhisthira A, Kim K-S, Bergey EJ; Prasad Paras N (2002) Nanochemistry: synthesis and characterization of multifunctional nanoclinics for biological applications. Chem Mater 14:3715–3721

Lewis JD, Destito G, Zjilstra A, Gonzalez MJ, Quigley J, Manchester M, Stuhlmann H (2006) Viral nanoparticles (VNPs) as tools for intravital vascular imaging. Nat Med 12:354–360

Li H, Li WX, Ding SW (2002) Induction and suppression of RNA silencing by an animal virus. Science 296:1319–1321

Lin T, Chen Z, Usha R, Stauffacher C, Dai J, Schmidt T, Johnson JE (1999) The refined crystal structure of cowpea mosaic virus at 2.8 A resolution. Virology 265:20–34

Liu L, Canizares MC, Monger W, Perrin Y, Tsakiris E, Porta C, Shariat N, Nicholson L, Lomonossoff GP (2005) Cowpea mosaic virus-based systems for the production of antigens and antibodies in plants. Vaccine 23:1788–1792

Lomonossoff G, Johnson J (1991) The synthesis and structure of comovirus capsids. Program Biophys Molec Biol 55:107–137

Lomonossoff G, Rohll J, Spall V, Maule A, Loveland J, Porta C, Usha R, Johnson J (1993) Insertion of foreign antigenic sites into the plant virus cowpea mosaic virus protein engineering. II. In: Goodenough P (ed) Proceedings of the Second AFRC Protein Engineering Conference. 1CPL Press, Newbury, UK, pp 30–138

Manayani DJ, Thomas D, Dryden KA, Reddy V, Siladi ME, Marlett JM, Rainey GJ, Pique ME, Scobie HM, Yeager M, Young JA, Manchester M, Schneemann A (2007) A viral nanoparticle with dual function as an anthrax antitoxin and vaccine. PLoS Pathog 3:1422–1431

Mandl S, Hix L, Andino R (2001) Preexisting immunity to poliovirus does not impair the efficacy of recombinant poliovirus vaccine vectors. J Virol 75:622–627

Mao C, Flynn CE, Hayhurst A, Sweeney RY, Qi J, Georgiou G, Iverson B, Belcher AM (2003) Viral assembly of oriented quantum dot nanowires. Proc Natl Acad Sci U S A 100:6946–6951

Mao C, Solis DJ, Reiss BD, Kottmann ST, Sweeney RY, Hayhurst A, Georgiou G, Iverson B, Belcher AM (2004) Virus-based toolkit for the directed synthesis of magnetic and semiconducting nanowires. Science 303:213–217

Martinez X, Regner M, Kovarik J, Zarei S, Hauser C, Lambert PH, Leclerc C, Siegrist CA (2003) CD4-independent protective cytotoxic T cells induced in early life by a non-replicative delivery system based on virus-like particles. Virology 305:428–435

Mechtcheriakova IA, Eldarov MA, Nicholson L, Shanks M, Skryabin KG, Lomonossoff GP (2005) The use of viral vectors to produce hepatitis B virus core particles in plants. J Virol Methods 131:10–15

Michel MC, Beck-Sickinger AG, Cox H, Doods HN, Herzog H, Larhammar D, Quirion R, Schwartz T, Westfall T (1998) XVI International Union of Pharmacology recommendations for the nomenclature of neuropeptide Y, peptide YY, and pancreatic polypeptide receptors. Pharmacol Rev 50:143–150

Morawski AM, Winter PM, Crowder KC, Caruthers SD, Fuhrhop RW, Scott MJ, Robertson JD, Abendschein DR, Lanza GM, Wickline SA (2004) Targeted nanoparticles for quantitative imaging of sparse molecular epitopes with MRI. Magn Reson Med 51:480–486

Phelps JP, Dang N, Rasochova L (2007) Inactivation and purification of cowpea mosaic virus-like particles displaying peptide antigens from *Bacillus anthracis*. J Virol Methods 141:146–153

Porta C, Spall VE, Lin T, Johnson JE, Lomonossoff GP (1996) The development of cowpea mosaic virus as a potential source of novel vaccines. Intervirology 39:79–84

Portney NG, Singh K, Chaudhary SK, Stephens J, Destito G, Schneemann A, Manchester M, Ozkan M (2005) Organic and Inorganic nanoparticle hybrids. Langmuir 21:2098–2103

Prasuhn DE Jr, Singh P, Strable E, Brown S, Manchester M, Finn MG (2008) Plasma clearance of bacteriophage Qbeta particles as a function of surface charge. J Am Chem Soc 130:1328–1334

Quintana A, Raczka E, Piehler L, Lee I, Myc A, Majoros I, Patri AK, Thomas T, Mule J, Baker JR Jr (2002) Design and function of a dendrimer-based therapeutic nanodevice targeted to tumor cells through the folate receptor. Pharm Res 19:1310–1316

Rae CS, Wei Khor I, Wang Q, Destito G, Gonzalez MJ, Singh P, Thomas DM, Estrada MN, Powell E, Finn MG, Manchester M (2005) Systemic trafficking of plant virus nanoparticles in mice via the oral route. Virology 343:224–235

Raja KS, Wang Q, Gonzalez MJ, Manchester M, Johnson JE, Finn MG (2003) Hybrid virus-polymer materials. 1. Synthesis and properties of PEG-decorated cowpea mosaic virus. Biomacromolecules 4:472–476

Rennermalm A, Li YH, Bohaufs L, Jarstrand C, Brauner A, Brennan FR, Flock JI (2001) Antibodies against a truncated *Staphylococcus aureus* fibronectin-binding protein protect against dissemination of infection in the rat. Vaccine 19:3376–3383

Ruedl C, Schwarz K, Jegerlehner A, Storni T, Manolova V, Bachmann MF (2005) Virus-like particles as carriers for T-cell epitopes: limited inhibition of T-cell priming by carrier-specific antibodies. J Virol 79:717–724

Schiappacassi M, Buratti E, D'Agaro P, Ciani L, Scodeller ES, Tisminetzky SG, Baralle FE (1997) V3 loop core region serotyping of HIV-1 infected patients using the FHV epitope presenting system. J Virol Methods 63:121–127

Schneemann A, Zhong W, Gallagher TM, Rueckert RR (1992) Maturation cleavage required for infectivity of a nodavirus. J Virol. 66:6728–6734

Scodeller EA, Tisminetzky SG, Porro F, Schiappacassi M, De Rossi A, Chiecco-Bianchi L, Baralle FE (1995) A new epitope presenting system displays a HIV-1 V3 loop sequence and induces neutralizing antibodies. Vaccine 13:1233–1239

Sedlik C, Saron M, Sarraseca J, Casal I, Leclerc C (1997) Recombinant parvovirus-like particles as an antigen carrier: a novel nonreplicating exogenous antigen to elicit protective antiviral cytotoxic T cells. Proc Natl Acad Sci U S A 94:7503–7508

Sen Gupta S, Kuzelka J, Singh P, Lewis WG, Manchester M, Finn MG (2005a) Accelerated bioorthogonal conjugation: a practical method for the ligation of diverse functional molecules to a polyvalent virus scaffold. Bioconjug Chem 16:1572–1579

Sen Gupta S, Raja KS, Kaltgrad E, Strable E, Finn MG (2005b) Virus-glycopolymer conjugates by copper(I) catalysis of atom transfer radical polymerization and azide-alkyne cycloaddition. Chem Commun (Camb) 4315–4317

Singh P, Gonzalez MJ, Manchester M (2005) Viruses and their uses in nanotechnology. Drug Dev Res 343:224–235

Singh P, Prasuhn D, Yeh RM, Destito G, Rae CS, Osborn K, Finn MG, Manchester M (2007) Bio-distribution, toxicity and pathology of cowpea mosaic virus nanoparticles in vivo. J Control Release 120:41–50

Soll RM, Dinger MC, Lundell I, Larhammer D, Beck-Sickinger AG (2001) Novel analogues of neuropeptide Y with a preference for the Y1-receptor. Eur J Biochem 268:2828–2837

Strable E, Johnson JE, Finn MG (2004) Natural nanochemical building blocks: Icosahedral virus particles organized by attached oligonucleotides. NanoLetters 4:1385–1389

Taylor K, Lin T, Porta C, Mosser A, Giesing H, Lomonossoff G, Johnson J (2000) Influence of three-dimensional structure on the immunogenicity of a peptide expressed on the surface of a plant virus. J Mol Recognit 13:71–82

Thiery R, Cozien J, Cabon J, Lamour F, Baud M, Schneemann A (2006) Induction of a protective immune response against viral nervous necrosis in the European sea bass Dicentrarchus labrax by using betanodavirus virus-like particles. J Virol 80:10201–1027

Wang Q, Kaltgrad E, Lin T, Johnson J, Finn M (2002a) Natural supramolecular building blocks: wild-type cowpea mosaic virus. Chem Biol 9:805–811

Wang Q, Lin T, Johnson J, Finn M (2002b) Natural supramolecular building blocks: cysteine-added mutants of cowpea mosaic virus. Chem Biol 9:813–819

Wang Q, Lin T, Tang L, Johnson J, Finn M (2002c) Icosahedral virus particles as addressable nanoscale building blocks. Angew Chem Int Ed 41:459–462

Wang Q, Chan TR, Hilgraf R, Fokin VV, Sharpless KB, Finn MG (2003) Bioconjugation by copper(I)-catalyzed azide-alkyne [3+2] cycloaddition. J Am Chem Soc 125:3192–3193

Yasawardene SG, Lomonossoff GP, Ramasamy R (2003) Expression, immunogenicity of malaria merozoite peptides displayed on the small coat protein of chimaeric cowpea mosaic virus. Indian J Med Res 118:115–124

Yewdell J, Anton LC, Bacik I, Schubert U, Snyder HL, Bennink JR (1999a) Generating MHC class I ligands from viral gene products. Immunol Rev 172:97–108

Yewdell JW, Norbury CC, Bennink JR (1999b) Mechanisms of exogenous antigen presentation by MHC class I molecules in vitro and in vivo: implications for generating CD8 T-cell responses to infectious agents, tumors, transplants and vaccines. Adv Immunol 73:1–77

Zinkernagel R (2002) On cross-priming of MHC class I-specific CTL: rule or exception? Eur J Immunol 32:2385–2392

Tumor Targeting Using Canine Parvovirus Nanoparticles

P. Singh (✉)

Contents

Introduction	124
Cancer and the Need for Novel Therapeutic Strategies	125
Transferrin Receptor Targeting for Tumor-Specific Delivery	125
Viruses in Materials Science	127
Challenges for Using VNPs for Biomedical Applications	127
Natural Tumor Specificity of Canine Parvovirus	128
Development of Methods for Production and Characterization of CPV-VLPs	132
Small Molecule Attachment to CPV	134
Natural Targeting of CPV and CPV-VLPs to Tumor Cells	136
Summary and Conclusions	136

Abstract Advances in genetics, proteomics and cellular and molecular biology are being integrated and translated to develop effective methods for the prevention and control of cancer. One such combined effort is to create multifunctional nanodevices that will specifically recognize tumors and thus enable early diagnosis and provide targeted treatment of this disease. Viral particles are being considered for this purpose since they are inherently nanostructures with well-defined geometry and uniformity, ideal for displaying molecules in a precise spatial distribution at the nanoscale level and subject to greater structural control. Viruses are presumably the most efficient nanocontainer for cellular delivery as they have naturally evolved mechanisms for binding to and entering cells. Virus-based systems typically require genetic or chemical modification of their surfaces to achieve tumor-specific interactions. Interestingly, canine parvovirus (CPV) has a natural affinity for transferrin receptors (TfRs) (both of canine and human origin) and this property could be harnessed as TfRs are overexpressed by a variety of human tumor cells. Since TfR recognition relies on the CPV capsid protein, we envisioned the use of virus or its shells as tumor targeting agents. We observed that derivatization of CPV virus-like particles (VLPs) with dye

---

P. Singh
Division of Hematology and Oncology, Department of Medicine, Building 23 (Room 436A),
UCI Medical Center, 101 City Drive South, Orange, CA 92868, USA
e-mail: pratiks@uci.edu

molecules did not impair particle binding to TfRs or internalization into human tumor cells. Thus CPV-based VLPs with a natural tropism for TfRs hold great promise in the development of novel nanomaterial for delivery of a therapeutic and/or genetic cargo.

Abbreviations AAV: Adeno-associated virus; CEA: Carcinoembryonic antigen; CPMV: Cowpea mosaic virus; CPV: Canine parvovirus; DMSODimethyl sulfoxide; IL-1β: Interleukin-1 beta; IL-6: Interleukin-6; NHS: N-hydroxysuccinimide; SEC: Size-exclusion chromatography; Tf: Transferrin; TfR: Transferrin receptor; TNF: Tumor necrosis factor; VNP: Viral nanoparticles; VLP: Virus-like particle

Introduction

More than half of the deaths between ages 50 and 85 in the United States in 2003 were due to cancer or heart disease (National Cancer Institute 2007). Although treatments for cancer range from generalized chemotherapeutics to more novel cell-specific therapies, the problems of toxicity and inadequate efficacy are still significant. Thus, specific targeting of tumor cells is an important consideration when designing therapeutics for optimal cancer treatment. Advances in genetics, proteomics and cellular and molecular biology are being integrated and translated to develop effective methods for the prevention and control of cancer.

One such combined effort is to create multifunctional nanodevices that will specifically recognize tumors and thus enable early diagnosis and provide targeted treatment of this disease. Viral nanoparticles (VNPs) are being considered for this purpose since they are inherently nanostructures with well-defined geometry and uniformity, ideal for displaying molecules in a precise spatial distribution at the nanoscale level and subject to greater structural control than possible with inorganic or lipid-based materials. Furthermore, viruses are presumably the most efficient nanocontainer for cellular delivery, since they have naturally evolved mechanisms for binding to and entering cells. Unfortunately, virus-based systems typically require genetic or chemical modification of their surfaces to achieve tumor-specific interactions.

Interestingly, examples of viruses with natural affinity or tropism for tumor cells exist. First, since tumor cells are rapidly dividing they provide an appealing host for viruses that rely on robust cell division for their replication. Several viruses with natural tumor tropism exist, such as paramyxoviruses, adenoviruses, and parvovirus. A particular example is the canine parvovirus (CPV). CPV has a natural affinity for transferrin receptors (TfRs) (both canine and human origin) and this property could be harnessed, since TfRs are overexpressed by a variety of human tumor cells. Since TfR recognition relies on the CPV capsid protein, we have envisioned the use of virus shells as delivery agents. Accordingly, we have produced CPV virus-like particles (VLPs) in a baculovirus expression system and subsequently attached small molecules to them. Based on the natural tropism of CPV for TfRs, we believe that both the VLPs and the complete virions can be engineered into a novel nanomaterial designed to target tumors for delivery of a therapeutic and/or genetic cargo.

Cancer and the Need for Novel Therapeutic Strategies

Normally, cells only grow and divide in a highly regulated and controlled manner. This orderly process helps keep the body healthy. Sometimes, however, cells will undergo unregulated replication to form a mass of tissue, called a tumor (National Cancer Institute 2007). Benign tumors are not cancerous and are rarely a threat to life. Cells from such tumors do not spread to other parts of the body and the entire mass can often be surgically removed. In contrast, malignant tumors are considered to be cancerous in that the comprising cells are abnormal and will divide continuously. Moreover, the resultant mass can invade and damage nearby tissues and organs. In addition, cancer cells can break away from the original or a secondary tumor and enter the circulatory system to initiate tumor formation in other organs (metastasis).

There are several main types of cancer (National Cancer Institute 2007). Carcinoma begins in the skin or in tissues that line or cover internal organs. Sarcoma originates in bone, cartilage, fat, muscle, blood vessels, or other connective or supportive tissue. Leukemia starts in blood-forming tissue such as the bone marrow and causes large numbers of abnormal blood cells to be produced and enter the bloodstream. Both lymphoma and multiple myeloma commence in the cells of the immune system (National Cancer Institute 2007).

A recent survey (Surveillance Epidemiology and End Results [SEER] cancer statistics review, National Cancer Institute [NCI] [Ries et al. 2007]) of cancer incidence and survival showed that among individuals ranging in age from 2 to 76 years, there were more deaths due to cancer than to heart disease. By 2020, it is expected that cancer rates may increase by 50%, with up to 15 million deaths worldwide (WHO 2003). The American Cancer Society estimates that more than one million new cases of cancer will be diagnosed in the United States and that roughly half a million Americans will die from this disease per year (American Cancer Society 2006). The overall expenditure for cancer treatment is estimated to be close to 210 billion dollars per year. Despite this financial output, the 5-year survival rates remain abysmally low for cancers of the pancreas (4% survival), lung (15%), and liver (7%), as well as glioblastoma (5%), a common form of brain cancer (American Cancer Society 2006). Even prostate and breast cancers, which are highly amenable to treatment if diagnosed early, are still difficult to treat during the later stages of disease and are responsible for more than 60,000 deaths a year (Edwards et al. 2005).

Transferrin Receptor Targeting for Tumor-Specific Delivery

Tumors are morphologically abnormal and therefore various cell surface and extracellular-matrix proteins can be utilized as markers to distinguish tumor from normal tissue (Ruoslahti 2002). Such abnormalities provide an opportunity for selective destruction without significantly affecting normal tissue. Tumor cells and their vasculature are logical prime targets for suppressing cancer growth. Although tumor vasculature is unusually leaky (Baluk et al. 2005; Hashizume et al. 2000),

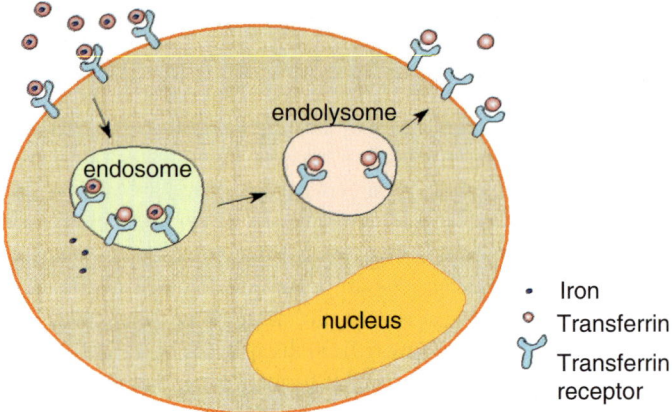

Fig. 1 Schematic representation of receptor-mediated endocytosis of transferrin. Iron-loaded transferrin binds to transferrin receptors followed by internalization of the complex into cells. The endosomal low pH causes the release of iron into the cytosol. Iron-deficient transferrin-receptor complex is either recycled to the cell surface or degraded

carrier systems use particles not exceeding 100 nm in diameter to ensure their tumor permeability and retention (Nagayasu et al. 1999; Tabata et al. 1998).

Internalization of nanoparticles into cells occurs by several mechanisms (Maxfield and McGraw 2004; Schmid 1997). Receptor-mediated endocytosis is a very specific and efficient process that is well-characterized and has been particularly harnessed for tumor-targeted delivery (Olivier 2005; Reddy 2005; Richter and Zhang 2005; Spicer and Harper 2005). The binding of the ligand to the receptor initiates the process of internalization, with the cargo eventually residing in the endosomes. The receptors are either recycled back to the cell surface after the ligand dissociates or the ligand–receptor complex is bound to lysosomes where they are degraded. The low pH environment of endosomes (5.0–6.0) favors the dissociation of most ligands (Maxfield and McGraw 2004) (e.g., see Fig. 1). The rationale for taking advantage of this system for specific delivery is that the receptors should be overexpressed on the cells to be targeted. Various targeting ligands including antibodies (Doronina et al. 2003; Spragg et al. 1997), transferrin (Qian et al. 2002), folic acid (Goren et al. 2000; Lu et al. 2004), oligopeptides and oligosaccharides (Zalipsky et al. 1997) are currently used.

Transferrin (Tf) is a circulatory iron-carrier protein that is in great demand, particularly during cellular growth and proliferation (Gomme et al. 2005). As iron is required by rapidly dividing cancerous cells, a significant upregulation of transferrin receptor expression is seen in a wide variety of tumor cells. Indeed, approximately 10^5 or more receptors per cell have been detected in several breast cancer cell lines (including MDA-MB-231) (Inoue et al. 1993), HeLa (human cervical carcinoma) (Bridges and Smith 1985), HT-29 (human colon carcinoma) (Becker et al. 2000), K562 (human erythroleukemia) (Bridges and Smith 1985; Sato et al. 2000) and pancreatic tumor cells (Ryschich et al. 2004). In contrast, very few if any

TfR molecules have been found on normal cells (Qian et al. 2002). Therefore tagging a drug or image contrast agent to transferrin for specific delivery to tumor cells has emerged as a promising strategy and is being widely explored for tumor-targeted delivery (Hogemann-Savellano et al. 2003; Qian et al. 2002).

Viruses in Materials Science

The use of materials derived from natural sources in materials science has allowed the harnessing of complex structures resulting from eons of evolutionary fine-tuning. A better understanding of the structure and function of viruses has revealed a collection of natural molecular assemblies and containers with a variety of shapes, sizes, stabilities, dynamic properties, and chemical reactivities (see reviews in Douglas and Young 2006; Singh et al. 2006). Viruses are increasingly being used in materials science, engineering and nanotechnology as tools and building blocks for electronics, chemistry and biomedical science. Viral particles are robust protein cages whose size is in the nanometer range (several under 100 nm). Their well-defined geometry and remarkable uniformity are ideal for nanoscale fabrication.

Virus capsids are versatile building blocks that can serve as scaffolds for producing nanomaterials (Brumfield et al. 2004; Wang et al. 2003). The atomic structures of many viruses have been resolved, allowing researchers to identify and if necessary modify amino acids in the viral capsid for bioconjugation. With the ease of using genetic manipulation to producing mutants displaying lysines or cysteines in the accessible virus capsid regions, it is feasible to chemically conjugate molecules to these amino acids regardless of whether they are located on the interior and exterior surface of the capsid. Moreover, as virus capsids are relatively rigid, molecules can be displayed in a precise spatial distribution at a nanoscale level. Already, bioconjugation to virus-based nanoparticles has been performed using commercially available homo- or hetero-bifunctional linkers and the lysines and/or cysteines of the viral capsid (Brown et al. 2002; Chatterji et al. 2004; Wang et al. 2002a, 2002b). More recently, we showed that an azide-alkyne "click" conjugation can be used for a specific and efficient linkage of ligands (including transferrin) to the virus capsid surface (Gupta et al. 2005; Wang et al. 2003). Together these techniques have provided many methods for conjugating various types of ligands to viral nanoparticles.

Challenges for Using VNPs for Biomedical Applications

The use of virus-based nanoparticles (VNPs) for biomedical purposes has several hurdles to overcome. Nevertheless, in view of the lack of efficient alternate methods, the benefits offered by this approach may outweigh the undesirable effects induced in vivo. In this regard, although both replication-competent and -defective human adenoviruses have been utilized for gene delivery purposes, these as well as their capsid components elicit the production of proinflammatory cytokines such as IL-6, IL-1β, and TNF-α and also elevate serum transaminases. As a result of this response, the treated individual experiences severe hepatotoxicity (Ben-Gary et al. 2002; Christ et al. 2000; Engler et al. 2004; Green et al. 2004; Higginbotham et al. 2002; Muruve et al. 2004). To counteract this effect, studies involving adenovirus or adeno-associated virus (AAV, a dependovirus belonging to the *Parvovirinae* subfamily) have demonstrated that a multivalent polymer coating applied to the virus capsid reduced toxicity and also presumably immunogenicity (Green et al. 2004; Lee et al. 2005). Nevertheless, because of its severe side effects, adenovirus is being displaced by recombinant AAV (rAAV) as the vector of choice for therapeutic gene transfer. Reasons for this substitution include the nonpathogenic nature of AAV and its derivatives and the virus's capability for efficient gene delivery to and sustained transgene expression in numerous tissues. Despite some initial successful applications in human clinical trials (Manno et al. 2003, 2006), certain rAAV constructs were found to be somewhat ineffective due to their neutralization by antibodies that recognize the progenitor's serotype and apparently preexist in a significant proportion of the human population (Moskalenko et al. 2000; Peden et al. 2004; Sun et al. 2003). To circumvent this problem, new rAAV vectors that are resistant to serum neutralization due to alteration of their capsid proteins by genetic (Maheshri et al. 2006) and chemical methods (Lee et al. 2005) have been produced. Another group within the *Parvovirinae* subfamily is the rodent parvoviruses. Unlike AAV, these viruses do not integrate into the cellular genome. Because rodent parvoviruses do not naturally infect humans and thus would not have previously triggered an immune response, they are being employed especially for tumor transduction. In addition, due to their preferential amplification in tumor cells and accompanying oncosuppressive and oncolytic traits, rodent parvoviruses are ideal for cancer therapy and providing transient expression of immunostimulatory genes to enhance tumor destruction (Geletneky et al. 2005; Rommelaere et al. 2005). Consequently, it is not surprising that patients injected with H1 rodent parvovirus have remained asymptomatic, despite the occurrence of viremia and seroconversion (LeCesne et al. 1993; Toolan et al. 1965). Thus, VNPs powered by the ease of genetic and chemical modifications continue to dominate as reagents for in vivo delivery of foreign genes (Young et al. 2006).

Natural Tumor Specificity of Canine Parvovirus

While the primary goal is enhanced efficacy through targeted tumor delivery, normal cells still must be spared from chemotherapeutic damage. Among the therapeutic devices now being created are nanoscale structures that are designed to perform as multifunctional diagnostic therapeutic agents. In this regard, viral particles are robust protein cages in the nanometer range and exhibit well-defined geometry and remarkable uniformity that are ideal for nanoscale fabrication. Accordingly, several viruses and VLP-based platforms have been utilized as nanocontainers for specific targeting applications (Douglas and Young 2006; Singh et al. 2006). However, these systems typically require modification of the virus surface by chemical or genetic means in order to attain tumor-specific delivery. Interestingly, there exists a virus, canine parvovirus (CPV), whose natural affinity for transferrin receptors (TfRs) on tumor cells could be exploited for nanobiotechnology applications. Our long-term goal is to develop CPV as a novel nanomaterial for tumor targeting.

The hypothesis underlying this research is that the natural TfR targeting of a CPV-based nanoparticle could be used to effectively target tumor cells and deliver a therapeutic payload, while avoiding interaction with normal cells. This premise is based on the following observations. First, CPV particles utilize TfRs for binding and entry into canine as well as human cells (Parker et al. 2001). While TfRs are barely detected on the surface of normal cells, they are several thousand-fold more abundant in a variety of actively dividing human tumor cells (Qian et al. 2002). Second, my work showed that the structural integrity of CPV-VLPs is not affected by chemical modifications such as the attachment of small molecules to them, and that the natural tropism of CPV for TfRs could be harnessed for a specific and targeted delivery of small molecules to human tumor cells (Singh et al. 2006). Third, CPV pathogenesis is host-restricted to canines. Although this virus can recognize TfRs on human cells, there is no evidence of CPV causing any disease in humans (Parker et al. 2001). Fourth, since activation of parvovirus promoters and replication of the parvovirus genome occurs during the S phase of the cell division cycle (Spelgelaere et al. 1991), a stage only entered into by actively dividing cells such as tumor cells, this dependency could be utilized to express foreign DNA packaged in CPV recombinants at specific sites. Last, in contrast to adenovirus-based platforms, parvoviruses do not induce a significant cytokine or toxic response in the host (Schlehofer et al. 1992).

CPV, a viral pathogen of canids (dogs), is a member of the family *Parvoviridae* (Muzyczka and Berns 2001). This infectious agent is an icosahedral (T=1), nonenveloped virus with a single-stranded DNA and an average diameter of 26 nm (Figs. 2, 3) (Tsao et al. 1991). The viral DNA encodes two polypeptides (VP1 and VP2) that are generated by alternative splicing of viral mRNA. A VP-3 subunit is formed only in full capsids by cleavage of 15–20 amino acids from the amino terminus of VP2. A full (DNA-containing) capsid is composed of 60 subunits, primarily VP2 (62 kDa) in nature, although a few VP1 and VP3 subunits are present. Empty capsids contain mostly VP2 subunits along with a minor amount of VP1 subunits (Tsao et al. 1991). Each subunit has a central jelly roll antiparallel β-barrel core with elaborate loops between the β-strands (Fig. 2) (Tsao et al. 1991). Generation of CPV- virus like particles (VLPs) in both mammalian cells and insects cells expressing only the coat protein gene has been described previously (Yuan and Parrish 2001). The transferrin receptor (TfR) on canine cells serves as a cellular receptor for the native CPV (Parker et al. 2001). Interestingly, infectious CPV particles were also found to bind and enter human cells via TfRs (Parker et al. 2001).

CPV virions encapsidate a genome of approximately 5.1 kb of single-stranded linear DNA of negative polarity, with palindromic sequences (shown as inverted

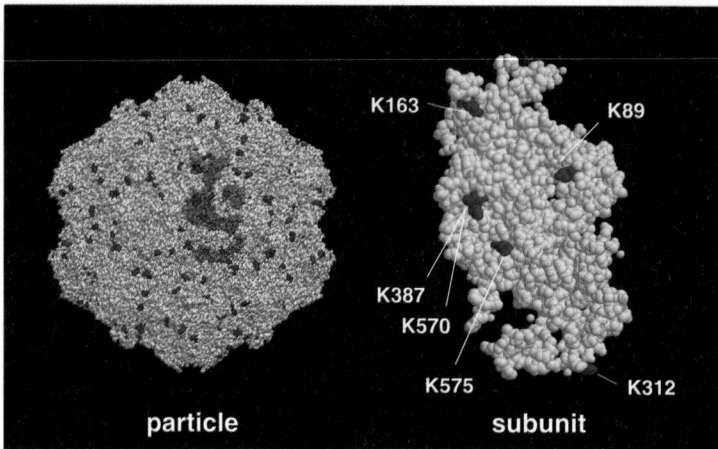

Fig. 2 Space-filling model of the canine parvovirus capsid. The model was generated using VMD software. The positions of lysines considered to accessible for chemical modifications in both the whole capsid (*left*) or an individual VP2 protein capsid subunit (*right*) are indicated. (Singh et al. 2006)

repeats; IR) at the termini that serve as self-priming origins of replication (Cotmore and Tattersall 1987; Muzyczka and Berns 2001; Reed et al. 1988). Within the genome there are two open reading frames (ORFs) coding for the nonstructural and structural viral proteins, driven by promoters P4 and P38 (Muzyczka and Berns 2001). The genomic organization and morphogenesis (typical of any parvovirus) of CPV is shown in Fig. 3. Expression of the nonstructural proteins, NS1 and NS2, is driven by the P4 promoter, whose activation is induced at the G1–S-phase transition (Spegelaere et al. 1991). Several oncoproteins have been shown to stimulate rodent parvoviral P4 promoter during G1-S-phase transition, resulting in NS1 production at a level high enough to initiate viral replication (Bashir et al. 2001; Deleu et al. 1999; Perros et al. 1995; Spegelaere et al. 1991). NS1 is a multifunctional protein essential for viral replication and promoter transactivation. It is also considered the major mediator of cytotoxicity in tumor cells (Caillet-Fauquet et al. 1990; Ozawa et al. 1988). NS2 is a 22-kDa protein that is essential for replication, virus production, nuclear egress of progeny virions and host-specific infection (Eichwald et al. 2002). In addition, NS2 seems to enhance NS1-mediated cytotoxicity to tumor cells (Brandenburger et al. 1990; Legrand et al. 1993). The P38 promoter is transactivated by the NS1 protein and regulates transcription of the capsid-coding genes, VP1 and VP2 (Lorson and Pintel 1997). The transcripts R1–3 and corresponding viral proteins that they encode are indicated in Fig. 3.

The life cycle of CPV is composed of following steps (Muzyczka and Berns 2001). Virus binds to the transferrin receptors and internalizes into the cell. Clathrin-coated vesicles mediate intracellular trafficking to the endosomes (Parker and Parrish 2000; Suikkanen et al. 2002). The virus replicates autonomously in the host cell nucleus. The transcripts, which translate to structural and nonstructural proteins, assemble in the cytoplasm close to nuclear envelope. The DNA is packaged into the capsid and the viral

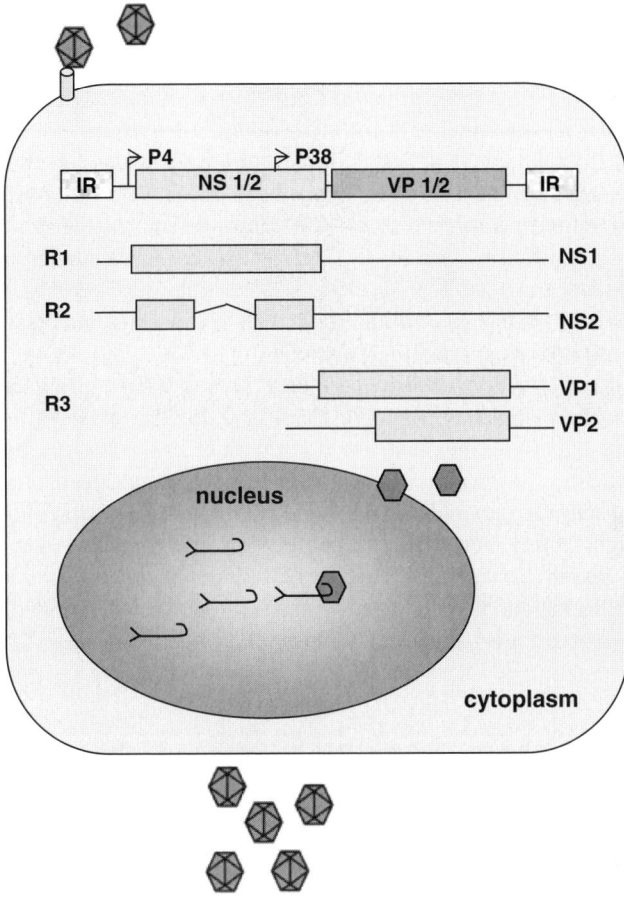

Fig. 3 Genomic organization and morphogenesis of CPV. Generation of replication proteins (NS 1/2) and capsid proteins (VP 1/2) are shown

particles are released in bursts (Eichwald et al. 2002). Interestingly, parvovirus infection of tumor cells results in oncolysis (Rommelaere and Cornelis 2001).

CPV infection in canine pups occurs by the oral-fecal route (Mar Vista Animal Medical Center n.d.). The virus in the throat and gastrointestinal tract reaches rapidly dividing groups of cells in the draining lymph nodes. Here the virus replicates and then virus particles enter the bloodstream to target organs containing rapidly dividing cells such as bone marrow and the delicate intestinal cells. Within the bone marrow, CPV is responsible for immunosuppression characterized by a drop in white blood cell count (Parrish 1995). The GI tract is where the heaviest damage occurs, causing diarrhea and susceptibility to bacterial infections. Fortunately, the entire immune system is not completely damaged, allowing neutralizing antibodies to be eventually produced that bind and inactivate the virus. Management of dehydration and prevention of secondary bacterial infections is effective in saving the animal (Mar Vista Animal Medical Center n.d.).

Viruses of the *Parvovirinae* subfamily have the ability to infect a variety of different vertebrates. Although the natural hosts of parvoviruses such as H1, minute virus of mice (MVM) and LuIII are rodents, these parvoviruses can also infect human cells (Rommelaere et al. 2005). In this regard, CPV was also found to generate infectious particles in HeLa cells (Parker et al. 2001). However, rodent or canine parvoviruses are not pathogenic or disease-causing in humans (Inoue et al. 1993; Parker et al. 2001; Rommelaere et al. 2005). Although viremia after human exposure to H1 rodent virus has been described, B19 (of the *Erythrovirus* genus) is the only virus of the *Parvovirinae* subfamily known to cause a human disease (Bloom and Young 2001; Brown et al. 1993).

In the context of cancer, parvoviruses have several unique characteristics. They were first isolated from human tumor tissue and for that reason were originally believed to be oncogenic (Rommelaere and Cornelis 1991). It was later discovered that oncogenic transformation of several human and rodent cells resulted in an enhanced capacity for parvoviral DNA amplification and gene expression and correlated with significantly increased susceptibility toward the parvoviral cytotoxicity in tumor cells. Rodent parvoviruses were found to have an oncosuppressive potential, inhibiting the formation of spontaneous and chemically or virally induced tumors in vivo and in vitro (Rommelaere and Cornelis 1991; Rommelaere et al. 2005).

Development of Methods for Production and Characterization of CPV-VLPs

CPV VLPs can be produced from baculovirus (Singh 2006), In addition, large quantities of CPV may be prepared using either the feline kidney cell line NLFK or CRFK (both feline kidney cell lines) as the host (Fig. 4). A typical virus production involved the following steps. Cells previously grown in McCoy/Leibovitz medium with 5% fetal calf serum (growth medium) were transferred to 16 roller bottles at a density of 210^4 cells/cm^2 in 100 ml growth medium per bottle and incubated at 37C. On the next day, the resultant monolayers in each bottle were washed with 10 ml Dulbecco's MEM + 0.1% BSA, overlaid with 5 ml of inoculum consisting of approximately 10^4 infectious center units of CPV per milliliter of infected cell lysate and incubated at 37C for 1 h. Afterward, the monolayers were covered with 100 ml growth medium and returned to the roller cabinet for 4 days. The infected monolayers and medium were then frozen and thawed twice. The combined contents of the roller bottles were adjusted to 0.25% NP-40, 5% Tris-HCl, pH 8.7, and 2% EDTA, mixed for 15 min at ambient temperature and stored at –20C until further processing.

For virus isolation, the above-mentioned infected cell lysate was first thawed and centrifuged at 9000 g for 30 min at 4C to remove cell debris. The clarified supernatant was then adjusted to 3.4% polyethylene glycol 8000 and 0.2 M NaCl and stirred

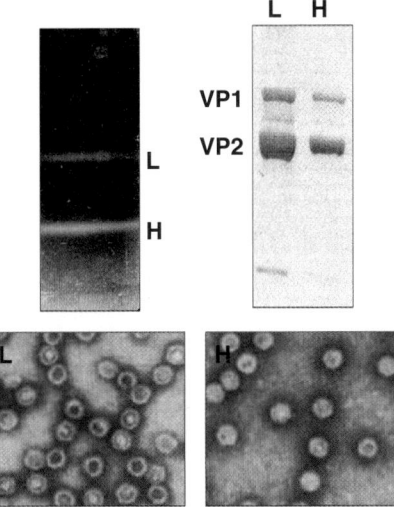

Fig. 4 CPV characterization. *Top left* Sucrose gradient (10%–40%) centrifugation of CPV particles. *L* light or empty virus particles; *H* heavy or full infectious virus particles. *Top right* SDS-PAGE analyses of L and H particles show the presence of both VP1 (84 kDa) and VP2 (62 kDa) polypeptides. *Bottom* Transmission electron micrograph of light (*left*) and heavy (*right*) particles showing empty and full capsids, respectively

overnight at 4C. The solution was centrifuged again at 9000 *g* for 30 min at 4C. The pelleted virus was resuspended in 40 ml of resuspension buffer (0.1% N-lauroylsarcosine in 0.01 M Tris-HCL, pH 7.5), prior to the addition of an equal of volume of chloroform. After vigorous mixing, the solution was centrifuged at 10,000 *g* for 15 min. The resultant aqueous (top) fraction was layered over a cushion of 20% sucrose in 10 mM Tris-HCl, pH 7.5, in several tubes and centrifuged at 30,000 rpm for 4.5 h. The pellet was resuspended in 3 ml of 10 mM Tris-HCl, pH 7.5, and sonicated six times with 10-s pulses. The sample was then layered on top of a 10%–40% sucrose gradient in 10 mM Tris-HCl, pH 7.5, and centrifuged for 4.5 h at 30,000 rpm. The visible light (L, empty virus particle) top band and the heavy (H, full virus particle) bottom band (Fig. 5) were collected with a syringe and separately dialyzed against phosphate buffered saline (PBS, pH 7.5). The quantity of virus in each band was determined based on one absorbance unit at 260 or 280 nm being equivalent to 7.0 or 1.4 mg/ml of heavy or light virus particles, respectively. The purified virus is stored at –70C and is considered stable for several years. The yield from 16 roller bottles ranges between 1–1.5 and 3–4.5 mg for heavy and light virus particles, respectively, with usually the heavy infectious virus portion being 25%–40% of the total yield. It should be noted that each milligram of purified CPV corresponds to approximately 2×10^{14} virus particles (Siegl et al. 1985). Actively dividing cells seems to be an important factor for optimal virus growth and yield and virus production is typically better in NLFK than CRFK cells.

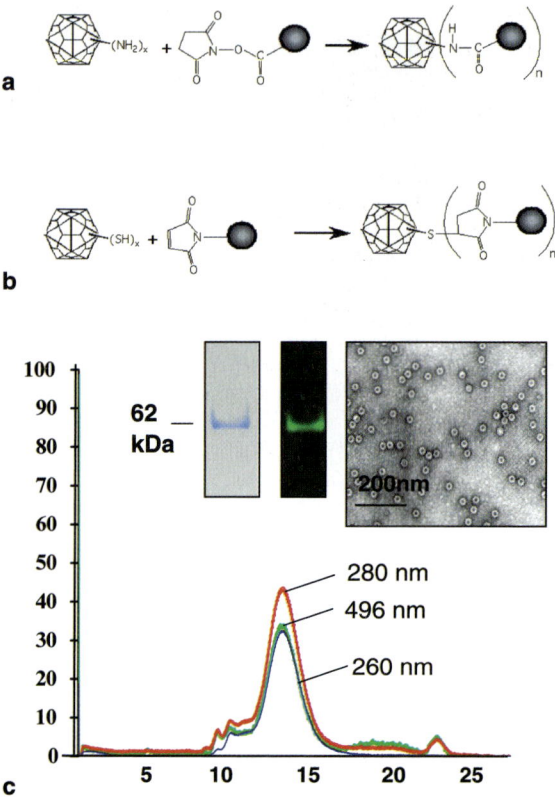

Fig. 5a–c Binding and internalization of CPV-VLPs labeled with OG-488 into tumor cell lines. Tumor cell lines HeLa (**a**), HT-29 (**b**), or MDA-MB-231 (**c**) were exposed to OG488-labeled CPV-VLPs. The cells were washed, fixed, and examined by confocal fluorescence microscopy for internalization of the labeled particles

Small Molecule Attachment to CPV

Molecules derivatized with NHS-esters can conjugate to exposed lysines on viral nanoparticles, while maleimide-derivatized molecules that are thiol selective can react and link with the cysteines (Fig. 6a, b). To evaluate whether CPV-VLPs could be efficiently derivatized by the same chemical methods used on other viral nanoparticles (Singh et al. 2006), the locations of surface lysines on CPV-VLPs were identified based upon a structural model of CPV using the radial distance and solvent accessibility surface area parameters in the VIPER database (Reddy et al. 2001). Using this model, two (maximum of six) lysines of the 20 present in each VP2 subunit, or 120 (maximal of 360) lysines per CPV-VLP particle could theoretically be accessible on the capsid surface for bioconjugation (Fig. 2). However, the reactivities are known to vary from their predicted accessibility due to the local

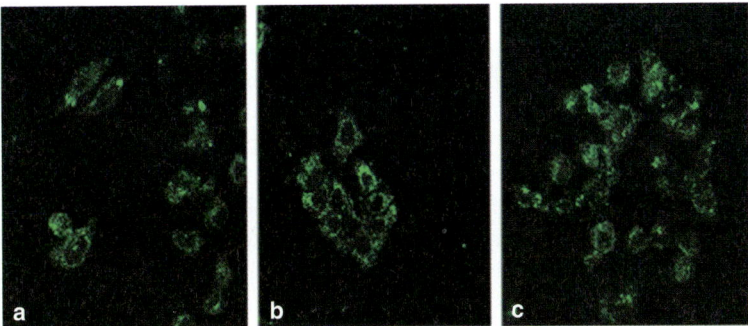

Fig. 6. a, b Chemical bioconjugation of CPV particles. Bioconjugation method for (**a**) lysines in virus capsid reacting with NHS-ester derivatized molecules or (**b**) thiols or cysteines in virus capsid reacting with maleimide-derivatized molecules. Molecule to be conjugated is shown as a filled circle. (**c**) Characterization of CPV-VLPs. Gel (*inset, left*): CPV-VLPs derivatized with OG-488 were subjected to SDS-PAGE and exposed to UV light (*inset, middle*) shows a 62-kDa band of VP2 protein. Electron micrograph (*inset, right*) of CPV-VLPs. Size exclusion chromatography (*bottom*) shows absorbance values recorded at 260, 280, and 496 nm, corresponding to peaks of viral nucleic acid, protein, and OG488 dye, respectively. The elution profile (in milliliters) from the column is shown on the x-axis

chemical environment of the lysine residue on the viral capsid surface (Wang et al. 2002a). Surface-accessible lysines on the capsid and on a single subunit of VP2 are depicted in a space-filling model in Fig. 2. In most cases, using 100 molar equivalents of OG-488 dye molecules per VP2 subunit in the VLP preparation, an average of 45 lysines/particle were addressed. Exposure to 200 molar equivalents of the dye per VP2 subunit resulted in an average of 100 derivatized lysines per particle. Further increases in the dye equivalents did not appear to enhance CPV-VLP labeling (data not shown). Dye-derivatized CPV-VLPs, when electrophoresed in a SDS-PAGE, appeared as a fluorescent band upon exposure to a UV-light source that migrated at 62 kDa (Fig. 6). Analyses of dye-labeled particles on the SEC at 496 nm revealed that the conjugate dye molecules were associated with the intact VLPs. TEM analyses of labeled VLPs showed an electron-dense core consistent with empty capsids (Fig. 6).

Recently we determined whether CPV-VLPs derivatized with OG-488 dye molecules showed cell binding and internalization characteristics similar to those exhibited by unmodified particles. To confirm the TfR specificity, the binding of OG-488-labeled CPV-VLPs to TRVb1 cells (expressing TfR) and TRVb cells (lacking or expressing very low levels of TfR) (McGraw et al. 1987) was investigated. Binding and internalization of dye-labeled CPV-VLPs was observed only in the TRVb1 cells, confirming that these two events are TfR-mediated. Thus the TfR-specific internalization of OG-488 labeled CPV-VLPs is similar to the native CPV-virions, in agreement with an earlier report (Parker et al. 2001). Since CPV-VLPs could withstand chemical conjugation and remain intact following purification, the potential of dye-labeled CPV-VLPs to target tumor cells, in addition to HeLa, was investigated. In this case, we examined binding and internalization of OG-488-labeled

CPV-VLPs into two other human tumor cell lines (HT-29 and MDA-MB231 cells) that are known to overexpress TfRs (Bridges and Smith 1985; Inoue et al. 1993). In both cases, similar to unlabeled particles in HeLa cells, the CPV-VLPs derivatized with OG-488 were invested within 2 h (Fig. 5).

Detailed analyses of the CPV capsid revealed that the Asn residues at positions 93 and 300 on the threefold spike are important in binding to the canine TfRs. Additionally, several amino acid residues in the shoulder region (Gly 299, Lys 387, Ala 300, Thr 301 and Val 316) also appear to play a role in binding (Parker et al. 1997). Based on CPV-capsid modeling (Fig. 2), the Lys residues at positions 89 and 312 are the most solvent-accessible and therefore more likely to be derivatized. In our bioconjugation experiments, attachment of dyes to the Lys 387 residue in some of the subunits could not be ruled out. However, the role of these residues in CPV binding specifically to TfRs has not yet been determined. Future studies will determine which residues are important reactive residues and which could be modified without interfering with TfR binding.

Natural Targeting of CPV and CPV-VLPs to Tumor Cells

To verify the predicted susceptibility of HeLa cells to CPV (Parker et al. 2001), glass cover slips in wells of a 24-well plate and were overlaid with growth medium containing 2×10^4 cells. After 24 h at 37C, the monolayers were infected with CPV at a multiplicity of 1 for 1 h at 37C, washed and overlaid with growth medium. After 24 h at 37C, the monolayers were washed with PBS, fixed with 4% paraformaldehyde, permeabilized in the presence of PBS containing 0.1% Triton X-100 and 1% BSA for 10 min and then exposed to a 1:3000 dilution of Alexa 594-labeled anti-NS1 antibody in PBS for 1 h at 25C. Afterward, the cover slips were washed three times with PBS and mounted on a glass slide with Immunomount medium. The cells were then examined using a fluorescent microscope for the presence of red fluorescence, indicative of CPVNS1 protein. NS1 protein was detected in the HeLa cell nuclei, demonstrating that CPV had entered and replicated in the cells.

CPV-VLPs were also shown to bind specifically to tumor cells and to enter cells in a TfR-dependent manner (Singh et al. 2006). Cells lacking the receptor did not bind to labeled VLPs, whereas cells expressing the receptor did. These results confirm the specificity of CPV-VLPs even after chemical derivatization. In addition, CPV-VLPs may be an interesting experimental tool to examine endocytosis via the transferrin receptor.

Summary and Conclusions

The elimination of cancer from an afflicted individual poses an enormous challenge to biomedical scientists. The National Cancer Institute has set a challenging goal of eliminating the suffering and death due to cancer by the year 2015. Various resources therefore are needed to fuel a multidisciplinary and coordinated effort in achieving this goal. Advances in genetic, molecular, proteomics and cellular events are being integrated and translated for developing an effective means of prevention, diagnosis and treatment of cancer. One such effort is to create targeted nanodevices with multifunctional capabilities to enable early diagnosis and offer superior treatment. Since the morphology of a tumor is quite distinct from that of normal cells, specific destruction of cancerous tissue should be achievable by focusing on its abnormalities. Strategies investigated for development of smart tissue-specific nanodevices include the use of nanospheres, quantum dots, dextrans, liposomes, antibodies, viral particles and dendrimers that are composed of targeting moieties along with cytotoxic drugs and/or imaging agents. In this regard, viral particles are inherently nanostructures with well-defined geometry and uniformity. As their capsids are relatively rigid, it is feasible to display molecules in a precise spatial distribution at a nanoscale level. It should be noted that achieving this level of control is not possible with inorganic or lipid materials. Furthermore, viruses are presumably the most efficient nanocontainer for cellular delivery, since they have naturally evolved mechanisms for binding and entering cells and utilizing the host's machinery for their survival and propagation. Thus canine parvovirus, with its natural tropism for transferrin receptors on tumor cells and its capability to deliver a bioconjugated cargo, holds great promise for development as a novel nanomaterial for tumor targeting.

References

American Cancer Society (2006) Cancer facts and figures, 2006. www.cancer.org/downloads/STT/CAFF2006PWSecured.pdf. American Cancer Society, Atlanta, GA. Cited 14 April 2008
Baluk P, Hashizume H, McDonald DM (2005) Cellular abnormalities of blood vessels as targets in cancer. Curr Opin Genet Dev 15:102–111
Bashir T, Rommelaere J, Cziepluch C (2001) In vivo accumulation of cyclin A and cellular replication factors in autonomous parvovirus minute virus of mice-associated replication bodies. J Virol 75:4394–4398
Becker A, Riefke B, Ebert B, et al (2000) Macromolecular contrast agents for optical imaging of tumors: comparison of indotricarbocyanine-labeled human serum albumin and transferrin. Photochem Photobiol 72:234–241
Ben-Gary H, McKinney RL, Rosengart T, et al (2002) Systemic interleukin-6 responses following administration of adenovirus gene transfer vectors to humans by different routes. Mol Ther 6:287–297
Bloom ME, Young NS (2001) Parvoviruses. In: Knipe DM Howley PM (eds) Fields virology, 4th edn., vol. 2. Lippincott Williams and Wilkins, Philadelphia, pp 2361–2379
Brandenburger A, Legendre D, Avalosse B, Rommelaere J (1990) NS-1 and NS-2 proteins may act synergistically in the cytopathogenicity of parvovirus MVMp. Virology 174:576–584
Bridges KR, Smith BR (1985) Discordance between transferrin receptor expression and susceptibility to lysis by natural killer cells. J Clin Invest 76:913–918
Brown KE, Anderson SM, Young NS (1993) Erythrocyte P antigen: cellular receptor for B19 parvovirus. Science 262:114–117
Brown WL, Mastico RA, Wu M, et al (2002) RNA bacteriophage capsid-mediated drug delivery and epitope presentation. Intervirology 45:371–380

Brumfield S, Willits D, Tang L, et al (2004) Heterologous expression of the modified coat protein of Cowpea chlorotic mottle bromovirus results in the assembly of protein cages with altered architectures and function. J Gen Virol 85:1049–1053

Caillet-Fauquet P, Perros M, Brandenburger A, et al (1990) Programmed killing of human cells by means of an inducible clone of parvoviral genes encoding non-structural proteins. EMBO J 9:2989–2995

Chatterji A, Ochoa W, Shamieh L, et al (2004) Chemical conjugation of heterologous proteins on the surface of cowpea mosaic virus. Bioconjug Chem 15:807–813

Christ M, Louis B, Stoeckel F, et al (2000) Modulation of the inflammatory properties and hepatotoxicity of recombinant adenovirus vectors by the viral E4 gene products. Hum Gene Ther 11:415–427

Cotmore SF, Tattersall P (1987) The autonomously replicating parvoviruses of vertebrates. Adv Virus Res 33:91–174

Deleu L, Pujol A, Faisst S, Rommelaere J (1999) Activation of promoter P4 of the autonomous parvovirus minute virus of mice at early S phase is required for productive infection. J Virol 73:3877–3885

Doronina SO, Toki BE, Torgov MY, et al (2003) Development of potent monoclonal antibody auristatin conjugates for cancer therapy. Nat Biotechnol 21:778–784

Douglas T, Young M (2006) Viruses: making friends with old foes. Science 312:873–875

Edwards BK, Brown ML, Wingo PA, et al (2005) Annual report to the nation on the status of cancer, 1975–2002, featuring population-based trends in cancer treatment. J Natl Cancer Inst 97:1407–1427

Eichwald V, Daeffler L, Klein M,, et al (2002) The NS2 proteins of parvovirus minute virus of mice are required for efficient nuclear egress of progeny virions in mouse cells. J Virol 76:10307–10319

Engler H, Machemer T, Philopena J, et al (2004) Acute hepatotoxicity of oncolytic adenoviruses in mouse models is associated with expression of wild-type E1a and induction of TNF-alpha. Virology 328:52–61

Geletneky K, Herrero YCM, Rommelaere J, Schlehofer JR (2005) Oncolytic potential of rodent parvoviruses for cancer therapy in humans: a brief review. J Vet Med B Infect Dis Vet Public Health 52:327–330

Gomme PT, McCann KB, Bertolini J (2005) Transferrin: structure, function and potential therapeutic actions. Drug Discov Today 10:267–273

Goren D, Horowitz AT, Tzemach D, et al (2000) Nuclear delivery of doxorubicin via folate-targeted liposomes with bypass of multidrug-resistance efflux pump. Clin Cancer Res 6:1949–1957

Green NK, Herbert CW, Hale SJ, et al (2004) Extended plasma circulation time and decreased toxicity of polymer-coated adenovirus. Gene Ther 11:1256–1263

Gupta SS, Kuzelka J, Singh P, et al (2005) Accelerated bioorthogonal conjugation: a practical method for ligation of diverse functional molecules to a polyvalent virus scaffold. Bioconj Chem 16:1572–1579

Hashizume H, Baluk P, Morikawa S, et al (2000) Openings between defective endothelial cells explain tumor vessel leakiness. Am J Pathol 156:1363–1380

Higginbotham JN, Seth P, Blaese RM, Ramsey WJ (2002) The release of inflammatory cytokines from human peripheral blood mononuclear cells in vitro following exposure to adenovirus variants and capsid. Hum Gene Ther 13:129–141

Hogemann-Savellano D, Bos E, Blondet C, et al (2003) The transferrin receptor: a potential molecular imaging marker for human cancer. Neoplasia 5:495–506

Hueffer K, Parrish CR (2003) Parvovirus host range, cell tropism and evolution. Curr Opin Microbiol 6:392–398

Inoue T, Cavanaugh PG, Steck PA, et al (1993) Differences in transferrin response and numbers of transferrin receptors in rat and human mammary carcinoma lines of different metastatic potentials. J Cell Physiol 156:212–217

LeCesne A, Dupressoir T, Janin N, et al (1993) Intra-lesional administration of a live virus, parvovirus H-1(PVH-1) in cancer patients: a feasibility study. Proc Ann Meet Am Soc Clin Oncol 12:297

Lee GK, Maheshri N, Kaspar B, Schaffer DV (2005) PEG conjugation moderately protects adeno-associated viral vectors against antibody neutralization. Biotechnol Bioeng 92:24–34

Legrand C, Rommelaere J, Caillet-Fauquet P (1993) MVM(p) NS-2 protein expression is required with NS-1 for maximal cytotoxicity in human transformed cells. Virology 195:149–155

Lorson C, Pintel DJ (1997) Characterization of the minute virus of mice P38 core promoter elements. J Virol 71:6568–6575

Lu Y, Sega E, Leamon CP, Low PS (2004) Folate receptor-targeted immunotherapy of cancer: mechanism and therapeutic potential. Adv Drug Deliv Rev 56:1161–1176

Maheshri N, Koerber JT, Kaspar BK, Schaffer DV (2006) Directed evolution of adeno-associated virus yields enhanced gene delivery vectors. Nat Biotechnol 24:198–204

Manno CS, Chew AJ, Hutchison S, et al (2003) AAV-mediated factor IX gene transfer to skeletal muscle in patients with severe hemophilia B. Blood 101:2963–2972

Manno CS, Arruda VR, Pierce GF, et al (2006) Successful transduction of liver in hemophilia by AAV-Factor IX and limitations imposed by the host immune response. Nat Med 12:342–347

Mar Vista Animal Medical Center (n.d.) The canine parvovirus information center. http://www.marvistavet.com/html/body_canine_parvovirus.html. Cited 9 April 2008

Maxfield FR, McGraw TE (2004) Endocytic recycling. Nat Rev Mol Cell Biol 5:121–132

McGraw TE, Greenfield L, Maxfield FR (1987) Functional expression of the human transferrin receptor cDNA in Chinese hamster ovary cells deficient in endogenous transferrin receptor. J Cell Biol 105:207–214

Moskalenko M, Chen L, van Roey M, et al (2000) Epitope mapping of human anti-adeno-associated virus type 2 neutralizing antibodies: implications for gene therapy and virus structure. J Virol 74:1761–1766

Muruve DA, Cotter MJ, Zaiss AK, et al (2004) Helper-dependent adenovirus vectors elicit intact innate but attenuated adaptive host immune responses in vivo. J Virol 78:5966–5972

Muzyczka N, Berns KI (2001) Parvoviridae: the viruses and their replication. In: Knipe DM Howley PM (eds) Fields virology, 4th edn., vol. 2. Lippincott Williams and Wilkins, Philadelphia, pp 2327–2359

Nagayasu A, Uchiyama K, Kiwada H (1999) The size of liposomes: a factor which affects their targeting efficiency to tumors and therapeutic activity of liposomal antitumor drugs. Adv Drug Deliv Rev 40:75–87

National Cancer Institute (2007) Defining cancer. www.cancer.gov/cancertopics/what-is-cancer, NCI, Bethesda, MD. Cited 14 April 2008

Olivier JC (2005) Drug transport to brain with targeted nanoparticles. NeuroRx 2:108–119

Ozawa K, Ayub J, Kajigaya S, et al (1988) The gene encoding the nonstructural protein of B19 (human) parvovirus may be lethal in transfected cells. J Virol 62:2884–2889

Parker JS, Parrish CR (2000) Cellular uptake and infection by canine parvovirus involves rapid dynamin-regulated clathrin-mediated endocytosis, followed by slower intracellular trafficking. J Virol 74:1919–1930

Parker JS, Murphy WJ, Wang D, et al (2001) Canine and feline parvoviruses can use human or feline transferrin receptors to bind, enter, and infect cells. J Virol 75:3896–3902

Parker SF, Felzien LK, Perkins ND, et al (1997) Distinct domains of adenovirus E1A interact with specific cellular factors to differentially modulate human immunodeficiency virus transcription. J Virol 71:2004–2012

Parrish CR (1995) Pathogenesis of feline panleukopenia virus and canine parvovirus. Baillieres Clin Haematol 8:57–71

Peden CS, Burger C, Muzyczka N, Mandel RJ (2004) Circulating anti-wild-type adeno-associated virus type 2 (AAV2) antibodies inhibit recombinant AAV2 (rAAV2)-mediated, but not rAAV5-mediated, gene transfer in the brain. J Virol 78:6344–6359

Perros M, Deleu L, Vanacker JM, et al (1995) Upstream CREs participate in the basal activity of minute virus of mice promoter P4 and in its stimulation in ras-transformed cells. J Virol 69:5506–5515

Qian Z, Li H, Sun H, Ho K (2002) Targeted drug delivery via the transferrin receptor-mediated endocytosis pathway. Pharm Rev 54:561–587

Reddy LH (2005) Drug delivery to tumours: recent strategies. J Pharm Pharmacol 57:1231–1242

Reddy VS, Natarajan P, Okerberg B et al (2001) Virus Particle Explorer (VIPER), a website for virus capsid structures and their computational analyses. J Virol 75:11943–11947

Reed AP, Jones EV, Miller TJ (1988) Nucleotide sequence and genome organization of canine parvovirus. J Virol 62:266–276

Richter M, Zhang H (2005) Receptor-targeted cancer therapy. DNA Cell Biol 24:271–282

Ries LAG, Melbert D, Krapcho M, Mariotto A, Miller BA, Feuer EJ, Clegg L, Horner MJ, Howlader N, Eisner MP, Reichman M, Edwards BK (eds) SEER Cancer Statistics Review, 1975–2004, National Cancer Institute. Bethesda, MD. http://seer.cancer.gov/csr/1975_2004/, based on November 2006 SEER data submission, posted to the SEER web site, 2007. Cited 14 April 2008

Rommelaere J, Cornelis JJ (1991) Antineoplastic activity of parvoviruses. J Virol Methods 33:233–251

Rommelaere J, Cornelis JJ (2001) Autonomous parvoviruses. In: Hernaiz Driever P, Rabkin SD (eds) Replication-competent viruses for cancer therapy, vol. 22. Kargel, Basel, pp 100–129

Rommelaere J, Giese N, Cziepluch C, Cornelis JJ (2005) Parvoviruses as anticancer agents. In: Sinkovics JG, Horvath JC (eds) Marcel Dekkar, New York, pp 627–675

Ruoslahti E (2002) Specialization of tumour vasculature. Nat Rev Cancer 2:83–90

Ryschich E, Huszty G, Knaebel HP, et al (2004) Transferrin receptor is a marker of malignant phenotype in human pancreatic cancer and in neuroendocrine carcinoma of the pancreas. Eur J Cancer 40:1418–1422

Sato Y, Yamauchi N, Takahashi M, et al (2000) In vivo gene delivery to tumor cells by transferrin-streptavidin-DNA conjugate. FASEB J 14:2108–2118

Schlehofer JR, Rentrop M, Mannel DN (1992) Parvoviruses are inefficient in inducing interferon-beta, tumor necrosis factor-alpha, or interleukin-6 in mammalian cells. Med Microbiol Immunol (Berl) 181:153–164

Schmid SL (1997) Clathrin-coated vesicle formation and protein sorting: an integrated process. Annu Rev Biochem 66:511–548

Siegl G, Bates RC, Berns KI, et al (1985) Characteristics and taxonomy of Parvoviridae. Intervirology 23:61–73

Singh P, Destito G, Schneemann A, Manchester M (2006) Canine parvovirus-like particles, a novel nanomaterial for tumor targeting. J Nanobiotechnology 4:2

Spegelaere P, van Hille B, Spruyt N, et al (1991) Initiation of transcription from the minute virus of mice P4 promoter is stimulated in rat cells expressing a c-Ha-ras oncogene. J Virol 65:4919–4928

Spicer J, Harper P (2005) Targeted therapies for non-small cell lung cancer. Int J Clin Pract 59:1055–1062

Spragg DD, Alford DR, Greferath R, et al (1997) Immunotargeting of liposomes to activated vascular endothelial cells: a strategy for site-selective delivery in the cardiovascular system. Proc Natl Acad Sci U S A 94:8795–8800

Suikkanen S, Saajarvi K, Hirsimaki J, et al (2002) Role of recycling endosomes and lysosomes in dynein-dependent entry of canine parvovirus. J Virol 76:4401–4411

Sun JY, Anand-Jawa V, Chatterjee S, Wong KK (2003) Immune responses to adeno-associated virus and its recombinant vectors. Gene Ther 10:964–976

Tabata Y, Murakami Y, Ikada Y (1998) Tumor accumulation of poly(vinyl alcohol) of different sizes after intravenous injection. J Control Release 50:123–133

Toolan HW, Saunders EL, Southam CM, et al (1965) H-1 virus viremia in the human. Proc Soc Exp Biol Med 119:711–715

Tsao J, Chapman MS, Agbandje M, et al (1991) The three-dimensional structure of canine parvovirus and its functional implications. Science 251:1456–1464

Wang Q, Kaltgrad E, Lin T, et al (2002a) Natural supramolecular building blocks: wild-type cowpea mosaic virus. Chem Biol 9:805–811

Wang Q, Lin T, Johnson J, Finn M (2002b) Natural supramolecular building blocks: cysteine-added mutants of cowpea mosaic virus. Chem Biol 9:813–819

Wang Q, Chan TR, Hilgraf R, et al (2003) Bioconjugation by copper(I)-catalyzed azide-alkyne [3+2] cycloaddition. J Am Chem Soc 125:3192–3193

WHO (2003) World cancer report. World Health Organization, Geneva

Young LS, Searle PF, Onion D, Mautner V (2006) Viral gene therapy strategies: from basic science to clinical application. J Pathol 208:299–318

Yuan W, Parrish CR (2001) Canine parvovirus capsid assembly and differences in mammalian and insect cells. Virology 279:546–557

Zalipsky S, Mullah N, Harding JA, et al (1997) Poly(ethylene glycol)-grafted liposomes with oligopeptide or oligosaccharide ligands appended to the termini of the polymer chains. Bioconjug Chem 8:111–118

Index

A
Adeno-associated virus, 128
Adenovirus, 76, 128
Alkynes, 15
Antigenic peptides, 26
Archaea, 76
Asymmetric derivatization, 86
Azides, 15

B
Bacillus anthracis, 115
Bacteria, 76
Baculovirus expression, 105
Bioavailability, 116
Bioconjugation, 7, 11, 34, 127, 134
    azide-alkyne click, 127
    cysteine, 127
    lysine, 127

C
Cancer, 124, 125
    benign, 125
    malignant, 125
    metastasis, 125
    tumor, 125
Canine parvovirus (CPV), 123, 124, 128
    bioconjugation, 134
    genomic organization, 129, 131
    infection, 131
    production, 132
    purification, 133
    replication, 129
    tumor cell binding, 135
Capsid, 2
Carbon nanotubes, 60
Catalysts, 81

Cellular immune responses, 114
Chemical attachment strategies, 34
Chimeric virus technology, 27
Conductive networks, 45
Copper(I)-catalyzed azide-alkyne
    cycloaddition, 14
Cowpea chlorotic mottle virus (CCMV), 12,
    74, 98
Cowpea mosaic virus (CPMV), 5, 24, 59, 98
    crystals, 49
Cryoelectron microscopy, 34, 41
CTL, 114
Cysteine, 13

D
2D and 3D arrays, 47
3D arrays, 49
2D CPMV monolayer, 54
Debye parameter, 67
Dendrimers, 10
Dendritic cells, 114
DLVO theory, 64
Doxorubicin, 85
Dps, 72

E
Electroactive, 42
Electrokinetic radius, 67
Eukarya, 76

F
Ferritin(s), 2, 12, 72
Ferrocenes, 42
Flock house virus (FHV), 102
Functionalities, 87

G
Gating properties, 83

H
Heat-shock proteins, 2
Hsp, 74
Human rhinovirus, 27
Hyperthermophile, 78

I
Image reconstruction, 34, 41
Imaging, 82
Immobilization, 50
Iron oxides, 79

L
Layer-by-layer, 50
Liposomes, 97
Lysine, 13

M
Microwave, 62
Mineralization, 77
MRI, 62, 82
Multivalent, 85

N
Nanocages, 2
Nanogold, 61
Nanoshells, 61
Network, 60
Nudaurelia capensis ω virus (NωV), 12

P
Phage display, 80
Phage display library, 85
Polyoxometalates, 79
Polyvalency, 24
Polyvalent scaffolds, 9
Polyvalent, 27
PRD1, 76
Protein cage, 73, 74

Q
Qβ, 15
Quantum dots, 60, 62, 97
Quartz crystal microbalance with dissipation monitoring (QCMD), 51

R
Receptor mediated endocytosis, 126
Redox-active, 42
Relaxivity, 82
Rodent parvovirus, 128, 132

S
Scanning tunneling microscopy, 45
Self-assembled, 47
Self-assembly, 24
Smoluchowski approximation, 66
Stern layer, 67
STIV, 72
Sulfolobus, 72, 75

T
Targeting, 110
Tobacco mosaic virus (TMV), 12
Toxicity, 118
Transferrin, 123, 124, 126
    expression, 126
    receptor, 130
Tumors, 96, 110
Turnip yellow mosaic virus (TYMV), 12
Tyrosine, 13

U
Ultra-small superparamagnetic iron oxide, 97

V
Vaccine candidates, 34
Vaccine(s), 26, 27, 30, 113
Vascular endothelium, 109
Vault proteins, 2
Viologen, 42
Viral nanoparticles (VNPs), 124, 127
Virus, 2
Virus-like particles (VLPs), 2, 132

Y
Yellowstone National Park (YNP), 72

Z
Zeta potential, 59, 63

Current Topics in Microbiology and Immunology
Volumes published since 2002

Vol. 271: **Koehler, Theresa M. (Ed.):** Anthrax. 2002. 14 figs. X, 169 pp. ISBN 3-540-43497-6

Vol. 272: **Doerfler, Walter; Böhm, Petra (Eds.):** Adenoviruses: Model and Vectors in Virus-Host Interactions. Virion and Structure, Viral Replication, Host Cell Interactions. 2003. 63 figs., approx. 280 pp. ISBN 3-540-00154-9

Vol. 273: **Doerfler, Walter; Böhm, Petra (Eds.):** Adenoviruses: Model and Vectors in VirusHost Interactions. Immune System, Oncogenesis, Gene Therapy. 2004. 35 figs., approx. 280 pp. ISBN 3-540-06851-1

Vol. 274: **Workman, Jerry L. (Ed.):** Protein Complexes that Modify Chromatin. 2003. 38 figs., XII, 296 pp. ISBN 3-540-44208-1

Vol. 275: **Fan, Hung (Ed.):** Jaagsiekte Sheep Retrovirus and Lung Cancer. 2003. 63 figs., XII, 252 pp. ISBN 3-540-44096-3

Vol. 276: **Steinkasserer, Alexander (Ed.):** Dendritic Cells and Virus Infection. 2003. 24 figs., X, 296 pp. ISBN 3-540-44290-1

Vol. 277: **Rethwilm, Axel (Ed.):** Foamy Viruses. 2003. 40 figs., X, 214 pp. ISBN 3-540-44388-6

Vol. 278: **Salomon, Daniel R.; Wilson, Carolyn (Eds.):** Xenotransplantation. 2003. 22 figs., IX, 254 pp. ISBN 3-540-00210-3

Vol. 279: **Thomas, George; Sabatini, David; Hall, Michael N. (Eds.):** TOR. 2004. 49 figs., X, 364 pp. ISBN 3-540-00534X

Vol. 280: **Heber-Katz, Ellen (Ed.):** Regeneration: Stem Cells and Beyond. 2004. 42 figs., XII, 194 pp. ISBN 3-540-02238-4

Vol. 281: **Young, John A. T. (Ed.):** Cellular Factors Involved in Early Steps of Retroviral Replication. 2003. 21 figs., IX, 240 pp. ISBN 3-540-00844-6

Vol. 282: **Stenmark, Harald (Ed.):** Phosphoinositides in Subcellular Targeting and Enzyme Activation. 2003. 20 figs., X, 210 pp. ISBN 3-540-00950-7

Vol. 283: **Kawaoka, Yoshihiro (Ed.):** Biology of Negative Strand RNA Viruses: The Power of Reverse Genetics. 2004. 24 figs., IX, 350 pp. ISBN 3-540-40661-1

Vol. 284: **Harris, David (Ed.):** Mad Cow Disease and Related Spongiform Encephalopathies. 2004. 34 figs., IX, 219 pp. ISBN 3-540-20107-6

Vol. 285: **Marsh, Mark (Ed.):** Membrane Trafficking in Viral Replication. 2004. 19 figs., IX, 259 pp. ISBN 3-540-21430-5

Vol. 286: **Madshus, Inger H. (Ed.):** Signalling from Internalized Growth Factor Receptors. 2004. 19 figs., IX, 187 pp. ISBN 3-540-21038-5

Vol. 287: **Enjuanes, Luis (Ed.):** Coronavirus Replication and Reverse Genetics. 2005. 49 figs., XI, 257 pp. ISBN 3-540- 21494-1

Vol. 288: **Mahy, Brain W. J. (Ed.):** Foot-and-Mouth-Disease Virus. 2005. 16 figs., IX, 178 pp. ISBN 3-540-22419X

Vol. 289: **Griffin, Diane E. (Ed.):** Role of Apoptosis in Infection. 2005. 40 figs., IX, 294 pp. ISBN 3-540-23006-8

Vol. 290: **Singh, Harinder; Grosschedl, Rudolf (Eds.):** Molecular Analysis of B Lymphocyte Development and Activation. 2005. 28 figs., XI, 255 pp. ISBN 3-540-23090-4

Vol. 291: **Boquet, Patrice; Lemichez Emmanuel (Eds.):** Bacterial Virulence Factors and Rho GTPases. 2005. 28 figs., IX, 196 pp. ISBN 3-540-23865-4

Vol. 292: **Fu, Zhen F. (Ed.):** The World of Rhabdoviruses. 2005. 27 figs., X, 210 pp. ISBN 3-540-24011-X

Vol. 293: **Kyewski, Bruno; Suri-Payer, Elisabeth (Eds.):** CD4+CD25+ Regulatory T Cells: Origin, Function and Therapeutic Potential. 2005. 22 figs., XII, 332 pp. ISBN 3-540-24444-1

Vol. 294: **Caligaris-Cappio, Federico, Dalla Favera, Ricardo (Eds.):** Chronic Lymphocytic Leukemia. 2005. 25 figs., VIII, 187 pp. ISBN 3-540-25279-7

Vol. 295: **Sullivan, David J.; Krishna Sanjeew (Eds.):** Malaria: Drugs, Disease and Post-genomic Biology. 2005. 40 figs., XI, 446 pp. ISBN 3-540-25363-7

Vol. 296: **Oldstone, Michael B. A. (Ed.):** Molecular Mimicry: Infection Induced Autoimmune Disease. 2005. 28 figs., VIII, 167 pp. ISBN 3-540-25597-4

Vol. 297: **Langhorne, Jean (Ed.):** Immunology and Immunopathogenesis of Malaria. 2005. 8 figs., XII, 236 pp. ISBN 3-540-25718-7

Vol. 298: **Vivier, Eric; Colonna, Marco (Eds.):** Immunobiology of Natural Killer Cell Receptors. 2005. 27 figs., VIII, 286 pp. ISBN 3-540-26083-8

Vol. 299: **Domingo, Esteban (Ed.):** Quasispecies: Concept and Implications. 2006. 44 figs., XII, 401 pp. ISBN 3-540-26395-0

Vol. 300: **Wiertz, Emmanuel J.H.J.; Kikkert, Marjolein (Eds.):** Dislocation and Degradation of Proteins from the Endoplasmic Reticulum. 2006. 19 figs., VIII, 168 pp. ISBN 3-540-28006-5

Vol. 301: **Doerfler, Walter; Böhm, Petra (Eds.):** DNA Methylation: Basic Mechanisms. 2006. 24 figs., VIII, 324 pp. ISBN 3-540-29114-8

Vol. 302: **Robert N. Eisenman (Ed.):** The Myc/Max/Mad Transcription Factor Network. 2006. 28 figs., XII, 278 pp. ISBN 3-540-23968-5

Vol. 303: **Thomas E. Lane (Ed.):** Chemokines and Viral Infection. 2006. 14 figs. XII, 154 pp. ISBN 3-540-29207-1

Vol. 304: **Stanley A. Plotkin (Ed.):** Mass Vaccination: Global Aspects – Progress and Obstacles. 2006. 40 figs. X, 270 pp. ISBN 3-540-29382-5

Vol. 305: **Radbruch, Andreas; Lipsky, Peter E. (Eds.):** Current Concepts in Autoimmunity. 2006. 29 figs. IIX, 276 pp. ISBN 3-540-29713-8

Vol. 306: **William M. Shafer (Ed.):** Antimicrobial Peptides and Human Disease. 2006. 12 figs. XII, 262 pp. ISBN 3-540-29915-7

Vol. 307: **John L. Casey (Ed.):** Hepatitis Delta Virus. 2006. 22 figs. XII, 228 pp. ISBN 3-540-29801-0

Vol. 308: **Honjo, Tasuku; Melchers, Fritz (Eds.):** Gut-Associated Lymphoid Tissues. 2006. 24 figs. XII, 204 pp. ISBN 3-540-30656-0

Vol. 309: **Polly Roy (Ed.):** Reoviruses: Entry, Assembly and Morphogenesis. 2006. 43 figs. XX, 261 pp. ISBN 3-540-30772-9

Vol. 310: **Doerfler, Walter; Böhm, Petra (Eds.):** DNA Methylation: Development, Genetic Disease and Cancer. 2006. 25 figs. X, 284 pp. ISBN 3-540-31180-7

Vol. 311: **Pulendran, Bali; Ahmed, Rafi (Eds.):** From Innate Immunity to Immunological Memory. 2006. 13 figs. X, 177 pp. ISBN 3-540-32635-9

Vol. 312: **Boshoff, Chris; Weiss, Robin A. (Eds.):** Kaposi Sarcoma Herpesvirus: New Perspectives. 2006. 29 figs. XVI, 330 pp. ISBN 3-540-34343-1

Vol. 313: **Pandolfi, Pier P.; Vogt, Peter K. (Eds.):** Acute Promyelocytic Leukemia. 2007. 16 figs. VIII, 273 pp. ISBN 3-540-34592-2

Vol. 314: **Moody, Branch D. (Ed.):** T Cell Activation by CD1 and Lipid Antigens, 2007, 25 figs. VIII, 348 pp. ISBN 978-3-540-69510-3

Vol. 315: **Childs, James, E.; Mackenzie, John S.; Richt, Jürgen A. (Eds.):** Wildlife and Emerging Zoonotic Diseases: The Biology, Circumstances and Consequences of Cross-Species Transmission. 2007. 49 figs. VII, 524 pp. ISBN 978-3-540-70961-9

Vol. 316: **Pitha, Paula M. (Ed.):** Interferon: The 50th Anniversary. 2007. VII, 391 pp. ISBN 978-3-540-71328-9

Vol. 317: **Dessain, Scott K. (Ed.):** Human Antibody Therapeutics for Viral Disease. 2007. XI, 202 pp. ISBN 978-3-540-72144-4

Vol. 318: **Rodriguez, Moses (Ed.):** Advances in Multiple Sclerosis and Experimental Demyelinating Diseases. 2008. XIV, 376 pp. ISBN 978-3-540-73679-9

Vol. 319: **Manser, Tim (Ed.):** Specialization and Complementation of Humoral Immune Responses to Infection. 2008. XII, 174 pp. ISBN 978-3-540-73899-2

Vol. 320: **Paddison, Patrick J.; Vogt, Peter K. (Eds.):** RNA Interference. 2008. VIII, 273 pp. ISBN 978-3-540-75156-4

Vol. 321: **Beutler, Bruce (Ed.):** Immunology, Phenotype First: How Mutations Have Established New Principles and Pathways in Immunology. 2008. XIV, 221 pp. ISBN 978-3-540-75202-8

Vol. 322: **Romeo, Tony (Ed.):** Bacterial Biofilms. 2008. XII, 299. ISBN 978-3-540-75417-6

Vol. 323: **Tracy, Steven; Oberste, M. Steven; Drescher, Kristen M. (Eds.):** Group B Coxsackieviruses. 2008. ISBN 978-3-540-75545-6

Vol. 324: **Nomura, Tatsuji; Watanabe, Takeshi; Habu, Sonoko (Eds.):** Humanized Mice. 2008. ISBN 978-3-540-75646-0

Vol. 325: **Shenk, Thomas E.; Stinski, Mark F.; (Eds.):** Human Cytomegalovirus. 2008. ISBN 978-3-540-77348-1

Vol. 326: **Reddy, Anireddy S.N; Golovkin, Maxim (Eds.):** Nuclear pre-mRNA processing in plants. 2008. ISBN 978-3-540-76775-6

Vol. 327: **Manchester, Marianne; Steinmetz, Nicole F. (Eds.):** Viruses and Nanotechnology. 2009. ISBN 978-3-540-69376-5